Bachem-Werke Ba 349 "Natter"

David Myhra

Schiffer Military History
Atglen, PA

Book Design by Ian Robertson.

Copyright © 1999 by David Myhra.
Library of Congress Catalog Number: 99-66304.

All rights reserved. No part of this work may be reproduced or used in any forms or by any means – graphic, electronic or mechanical, including photocopying or information storage and retrieval systems – without written permission from the copyright holder.
"Schiffer," "Schiffer Publishing Ltd. & Design," and the "Design of pen and ink well" are registered trademarks of Schiffer Publishing, Ltd.

Printed in China.
ISBN: 0-7643-1032-1

We are interested in hearing from authors with book ideas on military topics.

Published by Schiffer Publishing Ltd. 4880 Lower Valley Road Atglen, PA 19310 USA Phone: (610) 593-1777 FAX: (610) 593-2002 E-mail: Schifferbk@aol.com. Visit our web site at: www.schifferbooks.com Please write for a free catalog. This book may be purchased from the publisher. Please include $3.95 postage. Try your bookstore first.	In Europe, Schiffer books are distributed by: Bushwood Books 6 Marksbury Road Kew Gardens Surrey TW9 4JF England Phone: 44 (0)208 392-8585 FAX: 44 (0)208 392-9876 E-mail: Bushwd@aol.com. Try your bookstore first.

Bachem Ba 349

The *Bachem Ba 349* [*BP-20*] "*Natter*"—meaning viper or snake—was, near war's end, one of the desperate attempts by *Reichsführer-SS Heinrich Himmler* to somehow stop American *B-17* "*Flying Fortresses*" in aerial bombing Germany back to the stone age. *Dipl.-Ing. Erich Bachem* of Waldsee/Württemberg (about 25 miles [40 kilometers] from Lake Constance in southern Germany) had proposed a semi-disposable flying machine carrying a battery of two dozen *Henschel Hs 217 R- 4M 73* mm rockets in its nose and code-named "*Anseon*" to *Oberst Siegfried Knemeyer*, leader of the *RLM's* Technical Department. He and others laughed *Bachem*, with his uninvited proposal, right out of the *RLM*. *Bachem* then sought the assistance of *General der Jagdflieger Adolf Galland*. Although supportive, his power in the *RLM* was waining, having been fired as General of the Fighters by *Hermann Göring*. *Galland* had been pretty much without a job until *Hitler* allowed to form a fighter group of so-called "*expertern*" in January 1945 flying *Me 262s* and known as *JV 44*. *Bachem's* proposed "*Natter*" project continued to be ignored, and instead the RLM may have hardened their resistance because *Bachem* had now sought to get it approved via nontraditional means, that is, seeking out the *General der Jagdflieger*. Now having failed twice, it was at this time that *Bachem* in desperation sought an audience with perhaps the single most feared man in the *Third Reich*...*Reichsführer-SS* (Schutzstaffel) *Heinrich Himmler*. *Bachem* was invited in to see *Himmler*. *Bachem* told the *Reichsführer* about his *Natter* and how it would be launched vertically via a metal tower, or a 70 foot high pole fresh-cut from a pine tree could serve, as well. This "*Natter*" would be powered by a single *HWK 109-509A2* bi-fuel liquid rocket engine producing 3,307 pounds [1,500 kilograms] thrust. Four *Schmidding 109-533* solid-fuel rocket boosters giving 1,102 pounds [500 kilograms] thrust each would provide additional assistance during lift-off. *HWK 509s* were widely available since the abandonment of the *Me 163*. Later, said *Bachem*, he'd like to try powering the "*Natter*" with a *BMW 003R* combined turbojet engine and a bi-fuel liquid rocket engine. *Himmler* listened. *Bachem* continued on, saying that the machine would be built entirely out of wood by former wood furniture makers. *Bachem* said that he had initially taken his proposal to the *RLM's* Technical Director *Oberst Siegfried Knemeyer*. But he and his powerful colleagues, such as *Hans Ants, Helmutt Schelp, Roluf Lucht* and *Gottfried Riedenbach*, laughed at him and told him to get out. *Dr.-Med. Siegfried Ruff* of the *DVL* believed that the *349's* pilot would survive the anticipated 2.2 times gravity he'd experience at take-off. This was based on his research at *DVL*. (*Dr. Ruff* was the same age as *Reimar Horten*, and he helped him build the *Ho 1* and *Ho 2* sailplanes when they were teenagers in Bonn.) After his rejection, *Bachem* continued, he went to see *Adolf Galland*. He liked the idea and said he would try to get *Knemeyer* to change his mind, but *Galland* was unsuccessful, too. *Himmler* bid good bye to *Bachem* and that he'd hear from him in a few days. Twenty-four hours after *Erich Bachem* had spoken to *Heinrich Himmler*, a caller from the *RLM* informed him that they had reconsidered his proposed *BP-20* interceptor. He now had immediate approval to begin work on the project with the *RLM* designation of *Ba 349*. It was now August 1944. The project was classified as

A *Ba 349A* camouflaged in *Light Blue* upper surface with a dense mottle of *Gray-Violet 21* as seen from its port side during horizontal flight. Scale model by *Jamie Davies*.

Geheime Reichs Sache, which was a more secret category than *GeKdos* meaning that only designers and construction personnel concerned with the weapon knew about it. This had happened about December 1944 when *Oberst Dr.-Ing. Halder* of the *Flak Entwicklung* in the *RLM* was fired by the *Reichsfuhrung SS. Oberst Halder* had been in charge of all *Flak*, rockets, and rocket propelled aircraft. When *Halder* was fired by the *SS*, they took over all responsibility for such weapons, and it was all entrusted to *SS Führer Dr.-Ing. Kammler*, including the "Natter" project. As of December 1944, the *RLM* had no further insight into development of the *Ba 349*.

Description:
Officially, the *Bachem Ba 349* was a HWK 509 rocket propelled fighter-interceptor aircraft which would destroy an enemy bomber with the least expenditure of effort. It was planned for the purpose of providing a defense for vital targets which were being attacked by large USAAF *B-17* and other bomber formations. It was to take-off vertically, and after making its attack was to be abandoned by the pilot, who would then land by parachute, as would the rocket unit of the interceptor. It was designed to be built entirely of wood in small wood-working shops for rapid production. Systems for take-off and landing were automatic, and this would reduce pilot training to a bare minimum. The major factors in achieving this aim were:

(1) Utilization of vertical rocket-assisted take-off, together with separate landing of the pilot and the power unit by parachute. In this way pilots need be taught only to fly and shoot, and the long period spent in learning to land and take-off would be eliminated.

(2) The *Ba 349* was built of wood without the use of glueing presses. Parts could be made in small wood-working shops throughout the *Third Reich* without interfering with existing aircraft production. It was stated by *Willy Fiedler*, partner in the *Bachem Werke*, that only 600 man-hours were required for the production of the airframe minus the *HWK 509* bi-fuel liquid rocket unit. Manufacturing the *HWK 509* was relatively simple compared to that of a *BMW 003* or *Jumo 004* turbojet engine. Other advantages *Erich Bachem* claimed for his project were a substantial savings in fuel and steel, the ability to take-off from small cleared spaces in the woods, ease of transport, camouflage, and the ability to reuse the fuselage and power unit. *Bachem* claimed that his *Ba 349* could be used as a ship-born fighter, too, but no details are available.

Airframe
The fuselage was of simple construction, consisting of wooden bulkheads and stringers, all covered by a thin plywood sheeting. It was a mid-wing monoplane of very low aspect ratio (3.33) with a fin and rudder disposed above and below the tailplane. It was not symmetrical, although it was intended to be flown the right way up or inverted with equal ease. The wings were of rectangular planform without sweep-back or dihedral, having a symmetrical section and parallel with the fuselage axis and 25 mm below it. The wing contained wooden spars, ribs, plywood skin, and were 11.75 ft [3.5 meters] in span. The wing section was NACA 0012, and they contained no ailerons. Vertical and horizontal stabilizers, as well as rudders and elevators, were of all wood construction. The vertical stabilizer extended above and below the fuselage, while the horizontal stabilizer was mounted high on the vertical stabilizer above the fuselage. The rudders operated conventionally, while the elevators were fitted with controls *(Siemens K-12 servomotors)* which

A *Bachem Ba 349A* "Natter" seen shortly after lift-off and just before it began its horizontal flight heading straight into an Allied bomber formation. In the *349B* version, of which only three are believed to have been built at Waldsee/Württemberg, the 4x*Schmidding 533* booster rockets were positioned on the rear fuselage so that their jet nozzles were parallel to the jet nozzle of the *HWK 509* engine. The "B" version carried more *T* and *C-Stoff* and thus was expected full thrust for 4 minutes 36 seconds compared to the "A" version which had a powered endurance of 2 minutes 23 seconds. Scale model by *Jamie Davies*.

operated together as elevators or differentially as ailerons. These dual purpose control panels are called "elevons."

The jettisonable nose section was composed of the stamped metal, honeycomb arrangement of rocket tubes covered with plywood skin. Aft of the nose section of the cockpit contained the rocket firing control, hood, nose section jettison control, and the rip chord which pulled out the tail parachute.

The armament was located in the nose of the fuselage. Aft of the armament was the cockpit, then the fuel tanks at the center of gravity, while the rear of the fuselage housed the HWK 509 bi-fuel liquid rocket engine, which exhausted at the aircraft's tail. On the outside of the fuselage at its tail end were attached four *Schmidding 533 SR34* solid-fuel booster rockets for assisting take-off. When the booster rockets were expended (after 12 seconds) they fell away thanks to explosive bolts developed by *Karl Butter* of *Ernst Heinkel AG*.

Tailplane

The tailplane was also rectangular in planform without dihedral or sweep back, and the chord was parallel with the fuselage axis and 137.7 mm above it. Two other tailplanes had been constructed, namely 207.5 mm and 273.7 mm above the fuselage axis. The tailplane's profile had a thickness ratio of 11 percent and the same section as the wing.

Take-off Rockets

The *Schmidding 533 SR34* diglycol-dinitrate solid-fuel booster rocket gave 26,455 pounds [12,000 kilograms] impulse/second, or about 2,205 pounds [1,000 kilograms] thrust for 12 seconds. They were held in place by explosive bolts. This booster rocket was of the coated type, with an internal star-section charge and used a 48 mm diameter nozzle in the summer and a 42 mm diameter nozzle in the winter. The nozzles were inclined outwards to keep the jet clear of the fuselage. As in other booster rockets, these nozzles were inclined at an angle to the general axis of the rocket so that the thrust line would pass through the center of gravity of this manned missile. The setting of these boosters was quite critical if stability was to be ensured. A great deal of difficulty was encountered in seeking to achieve an even balance of power output from the four individual boosters during the "*Natter's*" initial unmanned testing. What happened was that differential thrust encountered during the "*Natter's*" launching tended to throw it out of control even before leaving the launch pad. If it cleared the launch pad, many flew out of control immediately afterward.

Power Unit

The standard *HWK 109-509A2* bi-fuel liquid rocket engine as used in the *Messerschmitt Me 163* was used. This rocket engine used *T-Stoff* [hydrogen peroxide and water] and *C-Stoff* [hydrazine hydrate and methyl alcohol]. The fuel tank for *T-Stoff* held 119 gallons [450 liters] and the tank for *C-Stoff* held 66 gallons [250 liters]. This fuel was sufficient for 80 seconds of full thrust. The *HWK 509* produced 3,750 pounds at full thrust.

The fuel feed system was very similar to that used in the *Me 163, Me 263*, and the *DFS 228*. A portion of the peroxide is drawn off and decomposed to drive a turbine on a common shaft with two worm-type propellant pumps. The steam producer was a porcelain-lined pressure vessel containing a wire screen, on which were distributed pellets of calcium (or potassium) permanganate. Feeding a stream of peroxide over this catalytic agent resulted in violent decomposition into superheated steam and gaseous oxygen. These resultants were piped to the turbine nozzle and, after spinning the rotor, were exhausted through a rectangular waste nozzle below the fuselage.

The accessory section of the *HWK 509* also contained an electric starter on the end of the turbine shaft, centralized fuel-feed control box with linkage control to pilot's throttle, pressure regulator valve for the steam

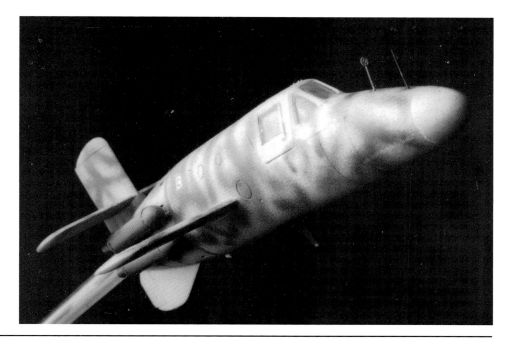

A *Ba 349A* still in a climbing mode as seen from its starboard side. *RLM* regulations did not allow for the German National Insignia (swastika) to be applied to disposable aircraft such as the "*Natter*." Scale model by *Jamie Davies*.

producer, automatic fuel cut-off valve, and a system filter. Thrust of the rocket power plant was transmitted to a built-up bulkhead in the fuselage by two tubular members which acted as the engine mount.

It was contemplated that a more simplified version of the HWK 109-509A2 would increase production and lower costs. Tests had shown that when the fuselage was landed by parachute the fuel remaining in the tanks and fuel lines was enough to cause an explosion, and the whole aircraft was lost.

Operating The HWK 509 Power Unit

Starting operation details presented here come from *Aviation 1/46* and consisted in initiating the pumping of the fluids, but did not start operation of the power plant. Movement of the control handle by the pilot in the cockpit to "idling" position energized the starter motor and opened the tank petcocks. The starter drove the pump at low power, causing the feed lines to become filled with propellants, but pressure developed at this time was not sufficient to overcome the valve setting in the main peroxide line to the steam producer. A bypass line fed back a small quantity of peroxide (from a pickup near the inlet valve) into the steam generator. After a few seconds of rotation the turbine was delivering enough power to the pumps to cause opening of the normal feed to the steam producer, and the bypass cut off.

Observing sufficient pressure registered on an indicator dial in the cockpit, the pilot moved the throttle to the "first" power setting. This resulted in the opening of three of the hydrate-methanol valves and three of the peroxide valves in the engine injection plate. Intermingling of the in-spraying fuels resulted in spontaneous combustion, thereafter continuous so long as propellants were fed into the combustion chamber.

Decomposition of the peroxide freed nascent oxygen, which burned with the alcohol to form the main source of heat. The violent reduction of the peroxide, under heat and catalytic action, released both thermal energy (adding to the velocity) and superheated steam (adding to the mass) of the exhaust.

Temperature of the engine was registered on a thermometer, visible to the pilot, calibrated from 300 to 1,000 degrees Centigrade. After the hydrate-methanol solution passed through the cooling jacket, it was returned to the fuel control center where its pressure was adjusted before injection into the combustion chamber. Movement of the control handle by the pilot into the "second" power stage opened an additional inlet for the hydrate-methanol and three additional inlets for the peroxide. The "high" power setting opened another hydrate-methanol (for a total of five) and six additional peroxide inlets (for a total of twelve). The HWK 509 came fully to life, operating with a loud roar and emitting a short blue-violet flame.

A system scavenging arrangement was incorporated which drained all propellants from the lines upon return of the control handle to the "off" position, and also shut off the tank's petcocks. A drain tube, connected to the cooling jacket with a pressure-operated valve in its line, served to exhaust the alcohol, while the peroxide vented through the turbine and waste nozzle.

Armament

The aircraft was fitted with a metal honeycomb for 24 *Henschel Hs 217 R4M Föhm* electrically-fired self stabilizing *73* mm rocket shells. *Bachem* planned to increase this number to as many as 48 *55* mm rocket shells. Prior to the surrender a *Ba 349* prototype was being developed which held 32 55 mm self stabilizing rocket shells in its nose. Alternative armament originally suggested was 2x*MK 108* 30 mm cannon, but this idea was dropped in favor of the *R4M*, which

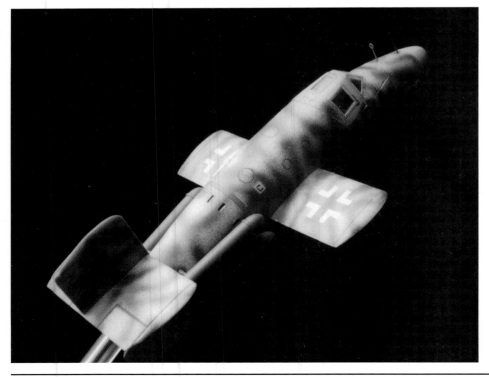

A starboard side view of a *Ba 349A*. On its nose cone can be seen the simple visual aiming sight device for getting a crude sighting prior to firing off its 24 nose-mounted *73* mm *R4M* rocket shells. Scale model by *Jamie Davies*.

achieved its stability through the use of fins which opened after leaving the guide rails instead of by rotation.

Dimensions

The following dimensions refer to the *Mark I* model of the *Ba 349*. Alterations were proposed and/or carried out throughout its short development time, such as increasing the wing span to 11.75 feet [3.6 meters]. Other dimensions include:

- Length, overall - 18 feet 7 inches [5.72 meters]
- Wing span - 10 feet 5 inches [3.20 meters]
- Tail span - 7 feet 5 inches [2.28 meters]
- Fuselage height w/o fins - 3 feet 8 inches [1.17 meters]
- Fuselage width - 2 feet 9 inches [0.90 meters]
- Fuselage height with fins - 6 feet 2 inches [1.90 meters]
- Wing chord - 3 feet 3 inches [1.00 meters]
- Wing loading, flight ready - 535 k/m²
- Wing loading, empty - 265 k/m²
- Wing area - 38.7 feet² [3.6 m²]

Weights

This weight summary also refers to the *Mark I* model of the *Ba 349*.

- Crew and equipment - 220 pounds [100 kilograms]
- Nose section - 66 pounds [30 kilograms]
- Offensive ammunition - 407 pounds [185 kilograms] @ 5 pounds 7 ounces [2.6 kilograms]
- Armored glass - 55 pounds [25 kilograms]
- Armor plate - 99 pounds [45 kilograms]
- Fuselage center section with wing - 242 pounds [110 kilograms]
- Fuel tanks and fuel - 1,463 pounds [665 kilograms]
- HWK 109-509A2 - 374 pounds [170 kilograms]
- Aircraft parachute - 66 pounds [30 kilograms]
- Fuselage rear section - 110 pounds [50 kilograms]
- Tail assembly - 154 pounds [70 kilograms]

Direct port side view of a *1101* armed with 2x*X-4* guided air-to-air missiles. Scale model by *Dan Johnson*.

- Total launching weight - 3,740 pounds [1,700 kilograms]
- Total launching weight - 4,840 pounds [2,200 kilograms] with 4x*Schmidding* 109-533 solid-fuel rocket boosters (4x275 pounds)

Performance

These figures refer to the *Mark I* model of the *Ba 349*.

- Maximum altitude - 39,300 pounds [12,000 kilometers]
- Climbing speed - 435 mph [700 km/h]
- Climbing rate - 37,000 feet per minute
- Acceleration at take-off - 2.2 times gravity slowing to 1.6 times gravity
- Acceleration, initial climb - 0.7 times gravity
- Vertical speed - 435 mph [700km/h] achieved in 5,000 feet [1,500 meters]
- Maximum horizontal speed - 16,500 feet [5,000 meters] - 620 mph [1,000 km/h]
- Normal cruising speed - 495 mph [800 km/h]
- Duration of flight - 2 minutes
- Radius of action at 39,300 feet [12,000 km] height - 12.5 miles [20 kilometers]

Miscellaneous

The Natter's cockpit controls were of the usual pattern, except that the control column could be hinged forward so as not to interfere with the pilot's exit from the aircraft. A few instruments, such as airspeed, height, rocket engine thrust, and directional gyro were installed. The aircraft's parachute was of the ribbon type and about 5 feet [1.5 meters] in diameter. It was stowed towards the rear end of the fuselage, with an attachment cable to the main bulkhead at which the motor is attached to the fuselage.

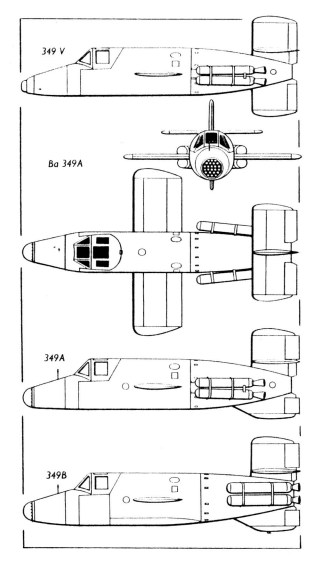

Above: A pen and ink 3-view drawing featuring the major physical differences between a *Ba 349V, 349A,* and a *Ba 349B*. The tail vertical assembly in the *349V* was changed considerably in the *349A* and featured an elongated ventral fin. Water-cooled fins in the *HWK 509's* exhaust orifice, designed by *Dipl.-Ing. Henry Bethpeder*, were to provide directional stability immediately after leaving the launch ramp. The mounting position of the 4x*Schmidding 533* booster rockets was changed, too. In the *349V* and *349A* the booster rocket's jet exited ahead of the *HWK 509* jet nozzle. With the *349B*, the boosters were placed further aft so that their 4xthrust nozzles were at the end of the fuselage...about equal to jet nozzle out of the *HWK 509* engine.

Operation

Initially, the launching apparatus known as a "*Lafette*" was employed. It consisted of two vertical poles imbedded in the ground and 49.25 feet [15 meters] high from ground level. Lugs on the wings ran in guides on the poles. A latter version consisted of a single pole 52.5 [16 meters] high in a concrete foundation 6.5 feet [2 meters] deep. Lugs on the ventral fuselage and fin ran in a channel on this pole, and additional support is provided by two flat strips parallel to the pole which were in contact with slide under the wings. The launching post could be rotated 360° about the foundation so that the cockpit of the aircraft was aiming towards any point of the compass, and was provided with a winch and tackle for setting the aircraft in place.

The aircraft was shot off with an initial acceleration of about 2.2 times gravity providing 37,000 feet per minute rate of climb. The booster rockets burned for about 12 seconds, at which time at a height of about 3,300 feet [1,000 meters] was reached. The expended rocket hulls would then drop off, and the aircraft continued on its climb at an acceleration of 0.7 times gravity until a speed of 435 mph [700 km/h] was reached. The pilot then throttled back and climbed at this steady rate of speed to intercept the bomber pack.

Under normal conditions, the *Ba 349* would be launched at the time the USAAF bomber formation was nearest to the starting point of the aircraft, and it would then be flown in an inclined path towards the bomber formation, attacking from the flanks. In guiding the *Ba 349* toward the bomber formation the pilot would be helped by *Flak* guns when available. A ground control radio was also under development. However, as the *Ba 349* neared the bomber formation it would attack with its *R4M* rocket shells, all of which were intended to be fired simultaneously from a range of 170 to 350 feet [50 to 100 meters]. After shooting down one or more bombers, the *Ba 349's* pilot would glide it down to about 9,000 feet [3,000 meters] if possible, jump out over a suitable and friendly area, and his speed would be further reduced. The pilot would jettison the nose of the fuselage, which then would fall away. He would release his safety harness and fold the control column forward. Folding the column forward would automatically release the *Ba 349's* tail parachute. This sudden deceleration due to the opening of the parachute, too, would propel the pilot forward and clear of the aircraft. Moments later he would open his own parachute and land in the normal way.

It was originally intended that the parachute would allow the *Ba 349* to land with a low sinking speed on its nose so that at least the power unit could be recovered and reused. However, it was discovered that residual fuel in the tanks and fuel lines exploded on landing and completely destroyed the aircraft. Therefore, the parachute's diameter was reduced in size to that necessary for only releasing the pilot. A method was being developed to separate the *HWK 509* rocket engine from the tanks and land the engine by parachute to be reused.

Modifications

Numerous small alterations were introduced in the *Ba 349* machine as development progressed. It is thought that 15 *Mark I* versions were produced. The *Mark II*, of which four aircraft of this type were found at St. Leonard, differed in that the fuselage was about 12 inches [30 centimeters] longer. In the *Mark III* the wing was to be situated further aft on the fuselage and was

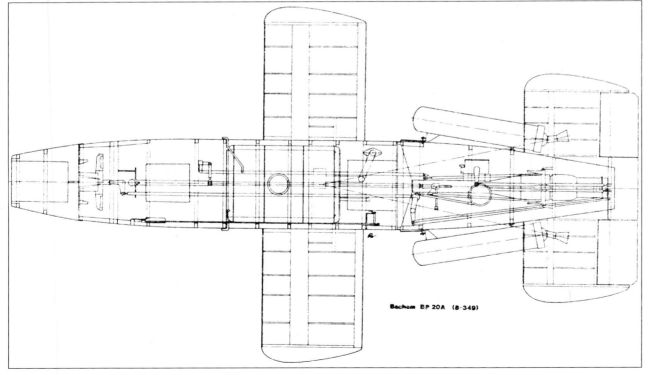

A pen and ink drawing of the general planform of a *Ba 349A* with internal details of its *T-Stoff* fuel tank (its *C-Stoff* fuel tank was carried directly beneath the *T-Stoff* fuel tank), placement of the *HWK 509* bi-fuel liquid rocket engine, and elevators on the horizontal stabilizer. There were no control surfaces on the wings.

removable for ease of transportation. For assembly, the *Mark III's* wing could be pushed right through an opening in the fuselage and secured with bolts. On this version the tailplane was to have been mounted at the top of the fin.

An important modification was the introduction of small horizontal surfaces coupled to the elevators into the exhaust/jet of the *HWK 509* rocket engine. The deflection of the jet by these surfaces gave control over the aircraft in the initial period before sufficient speed had been reached to enable the aerodynamic control surfaces to be effective. These jet-deflecting surfaces were filled with water and connected to a tank containing water. This brief cooling allowed them an operational life of 25 seconds before being burned off.

Willy Fiedler claimed during interrogation on 16 May 1945 that about the time he was active on the *Ba 349* project under *Waffen-SS General Wolff*, that the *SS* also had him working on a one way manned version of the *Fieseler Fi 103 V1* "buzz bomb" known as the *Fi 103R*. According to *Fiedler*, he personally believed that a manned *V1*, a kamikaze or suicide type aircraft, was not a very good idea, but that the *Waffen-SS* claimed to have 1,000s of *SS* volunteers for such suicide missions piloting a manned *V1*. *Fiedler* stated that a prototype had been flown and landed on skids. It had been launched from a *Heinkel He 111* parent aircraft. He had flown it only as a glider, but others had flown it under power. Noise surveys had shown that the noise of its intermittent pulse-jet engine was unbearable except in locations forward of the air intake. In addition, according to *Fiedler*, the *SS* said that they could immediately call up 100 volunteers to ram *Ba 349s* into *B-17* bombers. If the volunteer could bail out prior to the actual contact so much the better, because that pilot could take another "*Natter*" up and ram again—but the SS believed that very few of their volunteers would survive.

Flight Testing

Erich Bachem presented a proposal for his vertical launched *BP-20* interceptor in August 1944. At that time *Bachem* referred to it as the *BP-20*. It received a near unanimous rejection. In September 1944 *Bachem's BP-20* received the full support of *Reichsführer-SS Heinrich Himmler*, and the *RLM* followed immediately. All the initial experimental aircraft

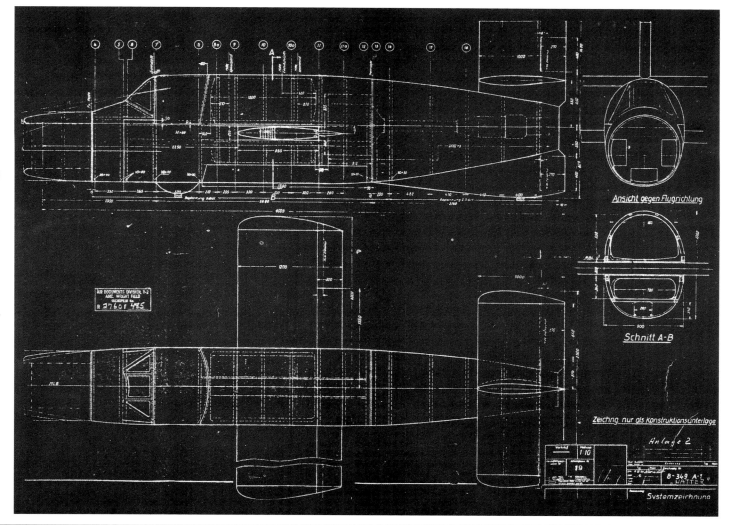

A pen and ink drawing of the *Ba 349A* port side fuselage, nose, location of the wing's main spar through the fuselage, and dorsal fuselage.

BP-20 Prototypes [M] Constructed and Ba 349 Serial Production*

BP-20 Prototypes [M] Constructed and Ba 349 Serial Production*

01. **M-1** - [BP-20] - First prototype constructed. Used to make flight tests after being towed to altitude by *DFS' He 111 H-6* (*DG+RN*). First unpowered manned flight made by *Hans Zübert* on 22 December 1944.
02. **M-2** - [BP-20] - Towed flight with fixed *Klemm Kl-35* tricycle landing gear by pilot *Hans Zübert*.
03. **M-3a** - [BP-20] - Towed flight. Pilot *Hans Zübert*.
04. **M-3b** - [BP-20] - Rebuilt *M-3a*. Towed flight by pilot *Hans Zübert*.
05. **M-4** - [BP-20] - Towed flight with separation/parachute trials.
06. **M-5** - [BP-20] - Towed flight with separation/parachute trials.
07. **M-6** - [BP-20] - Towed flight with separation/parachute trials.
08. **M-7** - [BP-20] - Towed flight with separation/parachute trials.
09. **M-8a** - [BP-20] - Towed flight. Pilot *Hans Zübert*.
10. **M-8b** - [BP-20] - Towed flight. Pilot *Hans Zübert*.
11. **M-9** - [BP-20] - Unknown duties/use.
12. **M-10** - [BP-20] - Unknown duties/use.
13. **M-11** - [BP-20] - Unmanned power launch. Unknown duties.
14. **M-12** - [BP-20] - Unmanned power launch. Unknown duties.
15. **M-13** - [BP-20] - Unmanned power launch.
16. **M-14** - [BP-20] - Unmanned power launch.
17. **M-15** - [BP-20] - Redesigned vertical tail surfaces with lower portion taking on the appearance of an elongated ventral fin plus water-cooled control vanes in the rocket exhaust orifice. All other *BP-20s* built afterward incorporated these changes. Unmanned power launch from the metal frame tower via remote control.
18. **M-16** - [BP-20] - Unmanned power launch via remote controls.
19. **M-17** - [BP-20] - Unmanned power launch via remote controls to 8,202 feet [2,500 meters].
20. **M-18** - [BP-20] - Unmanned power launch.
21. **M-19** - [BP-20] - Unmanned power launch.
22. **M-20** - [BP-20] - Unmanned power launch.
23. **M-21** - [BP-20] - Unmanned power launch.
24. **M-22** - [BP-20] - Unmanned power launch to test parachute deployment.
25. **M-23** - [BP-20] - First manned power launch on 1 March 1945. Piloted by *Lothar Sieber*. Crashed, killing pilot.
26. **M-24** - [BP-20] - Unmanned test aircraft thought to be under construction as of 1 March 1945.
27. **M-25** - [BP-20] - Manned test aircraft thought to be under construction as of 1 March 1945.
28. **M-26** - [BP-20] - Unknown duties/use/disposition.
29. **M-27** - [BP-20] - Unknown duties/use/disposition.
30. **M-28** - [BP-20] - Unknown duties/use/disposition.
31. **M-29** - [BP-20] - Unknown duties/use/disposition.
32. **M-30** - [BP-20] - Unknown duties/use/disposition.
33. **M-31** - [BP-20] - Launched from a 26.5 foot high [8.0 meter] tower for parachute testing.
34. **M-32** - [BP-20] - Launched for gantry testing.
35. **M-33** - [BP-20] - Unknown duties/use. Thought to be under construction as of 1 March 1945.
36. **M-34** - [BP-20] - Unknown duties/use. Thought to be under construction as of 1 March 1945.
37. **349A** - First production aircraft under construction as of 1 March 1945.

*This is not an absolute or complete list. Record-keeping within the *RLM* and *Luftwaffe* circles in late 1944 and early 1945 had broken down and/or were set afire thanks to the Allied bombing offensive, along with most everything else as Germany's war-making ability was collapsing throughout the country...falling head-long to its unconditional surrender on 8 May 1945. For example, *Hans Zübert* told this author that there were two *M-8*s and that he test flew both: *8a* and *8b*. *Zübert* also recalled that around *Bachem Werke*, at least, the machine which *Lothar Sieber* had test flown and died in, was known as the *M-9*, not the *M-23*, as is widely reported. Furthermore, *Hans Zübert* said that *Sieber's* test machine had numerals 2+3 painted on the upper surfaces of the wings. However, on the under surfaces were pained numerals 1+4. So what is correct? Difficult to say, and this author will make no attempt here to sort it all out...perhaps it can never be, given the sorry state of record-keeping throughout Germany at that late hour of the war.

to be used in manned or unmanned flight testing were built at the *Bachem Werke* in Waldsee/Württemberg.

Scale models of the *BP-20* were wind tunnel tested: one at *DVL*-Berlin and another at *LFA's* high speed wind tunnel at Braunschweig. *Erich Bachem* told U.S. military interrogators post war that he never received an analysis of the results other than a statement from DVL that the aircraft tested with the *RLM* designation of *Ba 349* should have satisfactory flying qualities up to 684 mph [1,100 km/h] speed and stable up to 32,800 feet [10,000 meters]. The *Ba 349's HWK 509* bi-fuel rocket engine could propel this aircraft up in excess of 32,000 feet. The *Boeing B-17* bomber packs usually operated between 25,000 and 32,000 feet, so their altitude range was well within the range of the "*Natter*."

But the *LFA*-Braunschweig's report on the aerodynamics of the *Ba 349* and the purpose for which *Erich Bachem* and *Willy Fiedler* designed it is really not all that bad. This is a summary of the *LFA*-Braunschweig's report on the "*Natter*."

• For the wing alone no large adverse compressibility effects can be seen from the shapes of the lift curves.
• The angle of zero lift remains almost constant throughout the whole Mach Number range.
• The aircraft appears to be stable at all altitudes up to 10 kilometers [32,800 feet], and is stable up to speeds of 800 km/h [497 mph], after which it then becomes unstable. The instability is more pronounced at sea level and decreases with altitude. This means that the *Ba 349's* nose must be pushed down for acceleration from Mach 0.70 to Mach 0.80, and at an altitude of 10 kilometers the nose must be pulled up for the same reason.
• The tailplane decreases in effectiveness above Mach 0.70.
• It appears that the underside of the tailplane is being adversely affected by the wing wake, and a higher position should give a more uniform efficiency over the range of positive and negative elevator angles.
• The main finding is that the tailplane becomes ineffective at high speeds because of the wake from the wing and is insufficient for stability in its present position.

Flight tests began when the first aircraft was completed in November 1944. *Flugkapitän Hans Zacker* and some of his assistants from the *DFS* -Ainring were called in to run these tests. The *Horten* brothers loaned *Hans Zübert* one of their all wing aircraft test pilots who had considerable experience in piloting sail (unpowered) planes. *Professor Ruden* of *DFS* was also involved in the gliding tests. Stability and control characteristics of the *Ba 349* in a dive up to 435 mph [700 km/h] and at stall, which occurred at 124 mph [200 km/h] at 3,937 feet [1,200 kilometers] were found to be excellent. *Hans Zübert* told this author that in gliding tests, however, the *349* sank like a rock.

For the initial "*Natter*," unmanned tests with the programmed three-axis *LGW* guidance systems of the older *K-12* type were installed. However, *Bachem* had hoped that his "*Natters*" might be able to use the simpler fighter control *K-23* later. The flight test program, in order of sequence, was to consist of unmanned takeoffs (some with an automatic pilot fitted), towed and gliding tests, powered flight tests at altitude, and finally manned take-offs. The initial take-off tests were made at Heuberg near Sigmaringen using 4x*Schmidding* booster rockets in place of the *HWK 509* rocket engine. The towed and gliding tests (three in all) were made at Heuberg and were designed to explore the stability, control, and stalling characteristics. *Hans Zübert* told this author that the test *Bachem* aircraft were towed behind with an extendable boom, cable, or vertically with cable below the *DFS' Heinkel He 111*. Upon reaching an appropriate attitude for the tests, with the extendable boom, for example, he was able to maneuver it around by means of a swiveling socket mount. *Zübert* recalled to this author of making a total of four manned gliding tests in the early versions of the *BP-20*. During these initial tests the "*Natter*" showed signs of having insufficient wing lift, not so much because of the design itself, but it was thought more due to the position of the extendable boom to which the tiny interceptor was attached. Upon release from the *DFS' He 111*, *Zübert* described how one of these *BP-20s* became so unstable that he had to abandon it. It was completely destroyed upon impact with the ground. The next four flight tests involved testing the *349's* parachute recovery system. These tests did not go well for the *BP-20s*, because although they were released by parachute, the airframe was still heavily damaged when meeting the ground.

Zübert's sixth *BP-20* flight test was more conventional. It required the machine to be towed to altitude,

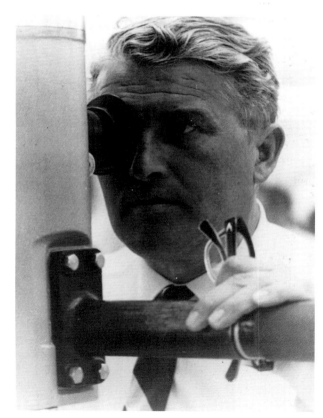

Dr.-Ing. Wernher von Braun post war seen peering through a periscope in a bunker at a missile launch site in the United States. It was *von Braun* who, in 1939, proposed to the *RLM* that they should consider the advantages of a vertically-launched interceptor.

released, and then landed in a grass field. A major modification had been made to the *BP-20* for this sixth test in that it had a crude non-retracting tricycle landing gear installed. The purpose was to retrieve the machine through a conventional landing, rather than parachute it down to earth as the others before had done. *Hans Zübert* told this author that this *BP-20* handled very well after release. He landed safely, although the *BP-20* was dropping like a rock and with a landing speed of over 120 mph!

On 18 December 1944 the first unmanned vertical launching of the *BP-20* was made with only 4x *Schmidding 533* solid-fuel rocket boosters for power. The near vertical lift-off was of short duration. The test machine failed even to clear the launch tower due to insufficient thrust. It became stuck about half way up the tower, and the wooden aircraft, with the 4x *Schmidding* rocket boosters burning feverishly, it was only moments before the entire machine was set afire with chunks of burning wood from its tail assembly falling down on its launch pad, which also had been set afire by the red-hot exhaust of the four *Schmidding* booster rockets. A second unmanned tower launch was tried and was somewhat more successful. This time the *BP-20* cleared the tower, but moments later fell back to the ground. *Bachem Werke* personnel persisted throughout December, January, and into February 1945. By early February improvements had been made, and launching a *BP-20* was becoming more and more successful. However, it appears that in late February 1945 the *SS* believed that *Bachem's* program was not progressing fast enough, according to *Hans Zübert*. It was at this time the *SS* ordered *Erich Bachem* to conduct a manned test flight with full *HWK 509* power plus its 4x *Schmidding* booster rockets. Officials at the *Bachem Werke* claimed that although the flight test program was showing continued success, it had not yet progressed to a manned flight test stage. Nevertheless, the impatient *SS* demanded that the flight-test program in fact be accelerated. The *SS* insisted *Erich Bachem* make a *BP-20* ready for a manned take-off. So it was that on 1 March 1945, pilot *Oberleutnant Lothar Sieber* in *Bachem's* test machine *M-23* and in a lying-on-his-back position pushed the several buttons to bring the *HWK 509* and *Schmidding 533s* alive. Bracing for the high-gravity forces he could expect to experience, higher than any human had perhaps ever experienced, the *BP-20's* one liquid and four booster rocket engines came to life. *Sieber's M-23* cleared the tower and continued ascending vertically to approximately 328 feet [100 meters] when ground personnel saw what appeared to be the cockpit canopy flying off. At this point the *BP-20* turned on its back, ascending to about 1,640 feet [500 meters] on a flight path at about 15 degrees from the horizontal. It then turned over into a nose dive, crashed to the ground, exploded, and was completely destroyed.

The exact cause of the crash and loss of its pilot *Oberleutnant Lothar Sieber* was not determined, but the actual summary of official findings (translated from the German) reads as follows:

Preliminary Brief Test Report on *M-23* First Vertical Launching of the Piloted Missile With Power Plant Take-off Assists - 1 March 1945

On 1 March 1945, pilot *Sieber* performed the first launching with a piloted missile from Heuberg.

The missile was set for full thrust. *Sieber* ignited the takeoff assists, and the missile left the mount satisfactory. After attaining an altitude of approximately 100 meters, the missile made a sharp turn in an upside-down position. This missile continued to climb, being inclined about 30%, whereby the hood dropped from the missile. After further climb to an altitude of approximately 1,500 meters, the power plant stopped approximately 15 seconds after launching. Thereupon the missile went into a power dive and hit the ground practically in a vertical descent, landing a few kilometers from the place of take-off, 32 seconds later. During the entire flight the pilot made no attempt to save himself.

The first vertical take-off of a piloted missile revealed the following:

1. It is assumed, and is highly probable, that the missile turned so rapidly on its back because the pilot, dazed by the take-off procedure, released the control stick arresting device, set forward degrees, whereby he pulled up the missile involuntarily, because of the accelerating forces.

2. Upon termination of the upside-down curve, the missile climbed in an inverted position at approximately 30 degrees. It might be assumed that

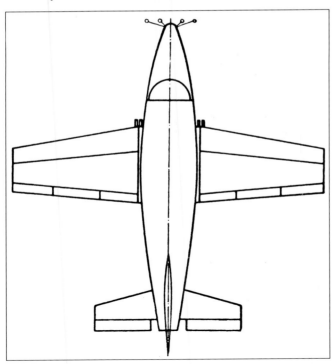

A pen and ink drawing of *Dr.-Ing. Wernher von Braun's* vertically-launching rocket powered target-defense proposal in his 6 June 1939 proposal to the *RLM*. It's planform was very conventional keeping with the accepted aircraft layout of the time, however, the *RLM* believed that there was no need for an interceptor. What country's aircraft could possibly penetrate *Luftwaffe* defenses and ever drop a bomb on the *Third Reich*?

the pilot's head was pushed against the hood by his own weight, so that the hood dropped off. While the hood lock had worked satisfactorily during a flight test with *M-8* up to a velocity of 600 km/h, it can nevertheless be stated that the lock was too weak to withstand the extraordinary stress. Hence the hood lock must be made stronger.

3. Since the hood and the attached head cushion fell off, the pilot struck his head against the back wall at high acceleration and probably became unconscious, so that he lost control over the missile temporarily. It is even possible that the pilot broke his neck at that moment, since his head was thrown back with such violence and since he slipped partly out of the missile, as far as the safety belt would permit.

4. Since during the entire flight the pilot made no attempt to save himself, it is to be assumed that during all this time he was completely dazed, or that he was not clear of his position in space after having shut off the power plant, so that he pulled out toward the vertical instead of remaining horizontal. Conclusions to be drawn from the above findings:

• The gravity load acting on the pilot and during take-off and possibly resulting confusion concerning the pilot's position in space require, as emphasized before, that the take-off process and the approximate target approach must be completely mechanized. Only the approach flight on enemy, i.e., firing approach (such as optical sighting of bombers), shall be left to the pilot.

• It is our opinion that another take-off with pilot is not to be made unless several missiles with automatics and without pilot have been flown satisfactorily. The first high-angle take-off with automatics and without pilot is set for approximately 10 March 1945.

• If, however, another high-angle start without automatics and with a pilot should be made, it could take place around 5 March, upon completion of the second manned missile *M-24*. Changes will consist in the heavier hood locking device and in locking the control stick during take-off in such a manner that the pilot cannot move the stick readily when suffering from shock, whereby a vertical ascent of the missile during the first 1,000 meters would be practically assured.

We suggest that the decision to this effect be reserved for the *SS-RHA* or *OKL*, as we do not approve a repetition of this test because disclosures with respect to further developments do not seem to warrant it.

Tests later made by *Bachem Werke* personnel showed that it was possible for the line of thrust of the 4x *Schmidding 533* booster rockets to deviate up to 2 degrees from the axis of the nozzle. Small rudders coupled to the elevators were added in the *HWK 509* exhaust/jet just aft of the nozzle in order to increase control immediately after take-off.

Several aviation historians claim that a second manned flight was made and that it also ended in disaster, however, this author is not aware of any evidence to support the claim of a second manned flight.

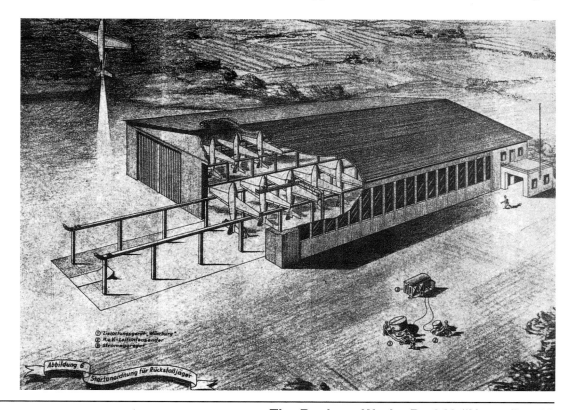

The launching complex suggested by *Wernher von Braun* in his second proposal to the *RLM* and dated 27 May 1941. *Von Braun* again proposed a vertical take-off interceptor. In this proposal he suggested his rocket-powered interceptor could be stored inside a hangar with a considerable amount of floor space. When the interceptors were needed they would be pulled out of the hangar to the launch area seen to the left of the illustration. It would be fired off as shown, then followed by another, and so on. It would appear that such a permanent launching complex would itself be subject to aerial bombing as was Peenemünde on the coast of the Baltic Sea. Again, the *RLM* declined.

Connection With The Japanese

Erich Bachem had been ordered by the *RLM* to give a complete set of drawings and details of his *Ba 349* to the Japanese. This was done, however, the submarine transporting this material to Japan never arrived and was presumed sunk and lost at sea. The Japanese had obtained a general description of the *Ba 349* and appeared to be highly interested in the project.

Number of Units Constructed

It is believed that between 30 to 36 *Ba 349* aircraft were built at the *Bachem Werke*- Waldsee/Württemberg. It is estimated by *Willy Fiedler* that another 30 were in various stages of completion at several other locations at war's end. Of the machines built at Waldsee/Württemberg, 18 were consumed on unmanned take-off flights. In addition, one aircraft was allowed to crash after its glide test, and another was destroyed in a manned test flight. One complete but broken-up *349* fuselage was found at *DFS*-Ainring. It had been used in gliding flight testing, being towed to altitude by a *DFS He 111*. Four new *Ba 349s* were purposely set afire at Waldsee/Württemberg, two set afire at Otztal, and four had been removed to St. Leonard by *Waffen- SS Oberleutant Flessner* prior to the American 44th Army moving into the area. *Willy Fiedler*, one of *Erich Bachem's* investors in the *Bachem Werke* and co-designer of the *BP-20*, claimed during interrogation post war that the *SS* had placed an order for ten *Ba 349s* with *Wolf Hirth Flugzeugbau* at *Nabern unter Teck* near Württemberg. *Wolf Hirth* was a well-known sailplane constructor and flyer, and he owned his own sailplane manufacturing company. At war's end only one *Ba 349* had been completed and delivered by *Wolf Hirth*. Shortly before Waldsee/Württemberg was occupied by the American Army, most of the *Bachem Werke* organization moved to Bad Werdershofen, and after a short time 13 *Waffen-SS* laborers led by *Waffen-SS Oberleutnant Flessner* moved on again, this time to *DFS*- facilities St. Leonard near Salzburg, Austria, with four new *Ba 349* and parts. It would be their last move. It appeared that they were waiting to be found there by the American Army's 44th Division with the hope that they could continue working on the *"Natter"* for the Americans for use against the Soviets. Those involved in the *"Natter"* project believed it was a state-of-the-art technology. Many of the former *Bachem Werke* officials were placed in a Prisoner of War camp, as were *Waffen-SS Oberleutnant Flessner* and his men.

The Number of *Ba 349s* Found At War's End

It is believed that at the end of the war only fourteen complete *Ba 349s* existed. In April 1945 ten flight-ready *Ba 349s* were taken to *Kirchheim on Teck* (near Stuttgart). None are known to have been fired off. They were found by the American Army's 44th Division. Two *Ba 349s* are reported to have been removed by *General George McDonald's* USAAF Intelligence-gathering teams, and at least one was transported to Freeman Field, Seymour, Indiana. In a report, head of USAAF Intelligence, *General McDonald*, stated the following items were packed and shipped to Headquarters, Freeman Field, Air Materiel Command, Seymour, Indiana:

- 2x *Ba 349* aircraft built by the *Bachem Werke*, Waldsee/Württemberg
- 2x *HWK 109-509A2* bi-fuel liquid rocket engines
- 4x *Schmidding 533 SR34* rocket booster engines
- 4x wings for a *Ba 349*
- 2x boxes of 73 mm *Henschel Hs 217 R4M* rocket shells
- 1x fuselage parachute for a *Ba 349*
- miscellaneous instruments, fittings, parts for a *Ba 349*
- 2x cabinets containing detailed drawings, parts lists, correspondence, and itemized modifications
- 1x compete set of drawings for the *HWK 509A2* bi-fuel liquid rocket engine
- 1x complete descriptive report of the *Ba 349* project
- A report on the examination of vertical take-off by *Dr.-Ing. K. Petrikat*, who did the aerodynamic work on the *Ba 349*.

According to *General McDonald's* report a *"Natter"* was obtained by the Red Army at Grüfenroda, Thüringen, where *Bachem* was getting ready to establish a manufacturing facility and had taken a factory-fresh *349* to be used as an example during construction of other machines.

A pen and ink drawing from 6 June 1939 of the port side of *von Braun's* vertically-launched target interceptor would have been powered by a single bi-fuel liquid rocket engine of his own design and producing an astonishing 22,050 pounds thrust. He estimated that his machine would reach 26,250 feet in 53 seconds.

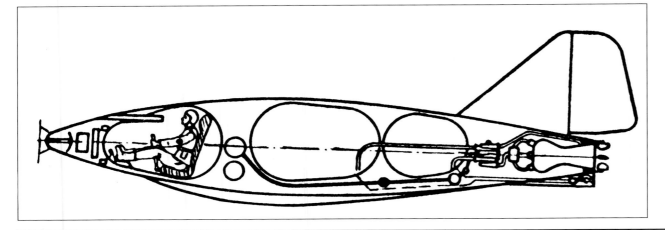

Bachem Ba 349 Surviving In The Year 2000

Only two *Ba 349s* are known to exist in the world. One is owned by the *Deutsches Museum*, Munich, Germany. It is beautifully restored, on display, and camouflaged in the paint scheme used on the *M-17*. The National Air and Space Museum, Washington, D.C., owns a *Ba 349A*. It is known by its foreign equipment number *T2-1011* and was one of the two *349A's* obtained from *DFS*-St. Leonard, Austria. It is currently in storage. This machine has never been flown. When it was taken to Freeman Field it was placed on static display about May 1946. Later it was transferred to Park Ridge near Chicago and was obtained by NASM in the early 1950s. It is unrestored and in storage at the museum's Restoration Center, Silver Hill, Maryland. It is not known to this author what became of the second *Ba 349A* obtained at St. Leonard and reported by *General McDonald* to have also been transported to Freeman Field. This author has a photo of a *349A* with a mottle camouflage...unlike the *T2-1011*, which had been painted a dark gray post-war. It was on public display at Douglas Airfield, Santa Monica, California, in the late summer of 1945. It has never been seen again. There is a persistent rumor that a *349A* was at RAF-Farnborough post-war, however, there are no official records of a *Ba 349* ever being there. This does not mean that one never was at RAF-Farnborough. Post-war *Captain Eric Brown* of the RAF-Farnborough in a letter to this author said that post-war he had gathered up the remains of the twin turbojet-powered *Horten Ho 9 V2* which crashed at Oranienburg 18 February 1945. *Brown* stated that he brought the center section to Farnborough, yet there are no known documents and/or photographs recording that the remains of the all-wing *Ho 9 V2* was ever brought to England!

This ends the story of *Erich Bachem* and *Willy Fiedler's BP-20* idea...a projectile midway between a guided missile and a manned-controlled fighter interceptor. It was size-wise the tiniest fighter interceptor in the *Luftwaffe's* inventory. The *Bachem Werke* is noteworthy for another reason...it enjoyed the shortest life span of any German aircraft manufacturing company during the *Third Reich*...Summer 1944 to Spring 1945.

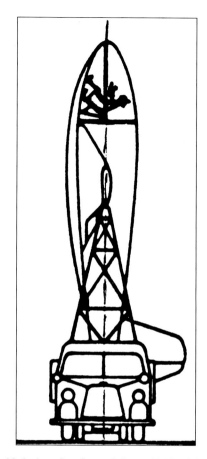

A pen and ink drawing from 6 June 1939 of the port side of *von Braun's* vertically-launched aircraft as it might have looked mounted between its two guide rails. *Erich Bachem* was a big fan of *von Braun*. It appears that *Dr.-Ing. von Braun's* vertically launched target interceptor ideas of June 1939 and May 1941 were the basis for *Erich Bachem's* "Natter" project of mid-1944.

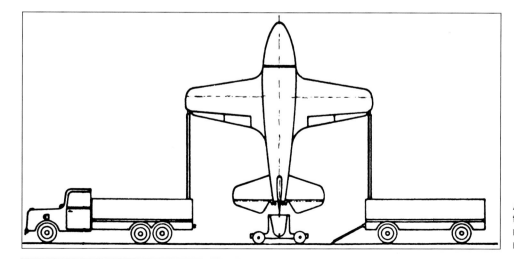

A pen and ink drawing from 6 June 1939 showing the general planform of *von Braun's* target defense interceptor. It would have been launched from two 20 foot high guide rails which really only held the machine steady prior to lift-off...similar as today's manned rockets.

The "Natter" was not Erich Bachem's first idea for a vertically launched interceptor. In the early 1940s when he was employed by the Gerhard Fieseler Werke, Bachem had presented his Fieseler Fi 166 to the RLM as a target defense interceptor. It was turned down because defensive aircraft were considered unnecessary within the German Luftwaffe. Shown here is Erich Bachem (left) with his family and their pet German Shepard dog. He left Fieseler in early 1943 to form his own company at Waldsee/Württemberg to manufacture wooden parts for air torpedoes, wooden control surfaces, and pieces for the V-1's fuselage. Fieseler Werke chief test pilot Willy Fiedler invested money in Bachem's aviation wood-working company and were partners even before the "Natter" project was started. People speak very highly of Erich Bachem. Willy Fiedler recalled that Bachem was a good person, good to work with and fun to be around. He enjoyed having people come to his house for dinner, even during the most difficult days of the war. He and his wife loved entertaining at their home in Kassel, and you often see photographs of Erich Bachem playing his accordion for their guests.

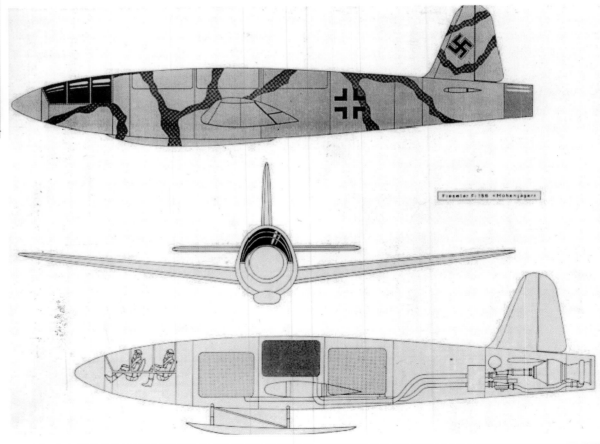

A pen and ink drawing of Erich Bachem's proposed 2-man bi-fuel liquid rocket-powered interceptor of the early 1940s. Bachem had been chief of Technical Development at the Gerhard Fieseler Werke and his design project was known as the Fieseler Fi 166.

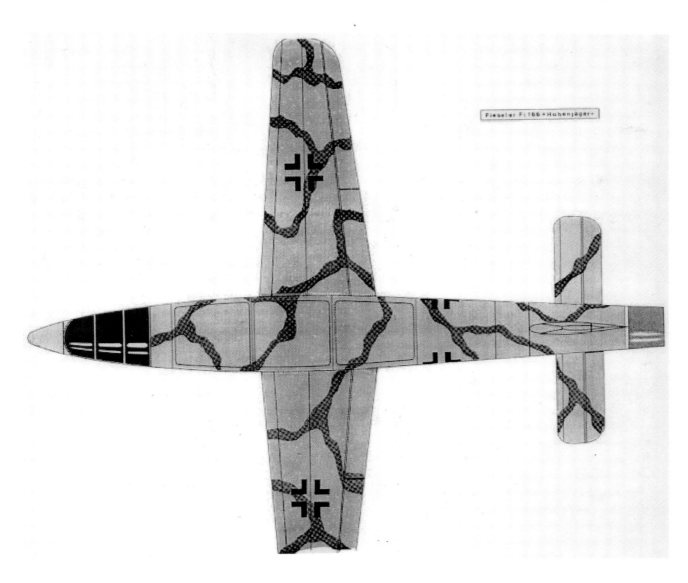

Above: A pen and ink drawing of the general planform of the *Fieseler Fi 166*. Former *Fieseler Werke* aircraft designer *Willy Fiedler* stated during his post-war interrogation by the U.S. Military that *Gerhard Fieseler* had sought to interest the *RLM* in his powered interceptor but that they were simply not interested in target defense machines at that time of the war.

Right: A pen and ink illustration of one version of the *Bachem/Fiedler Fi 166* known as *Design I*. It was to have been lifted-off on the back of one of *von Braun's A-4* rockets...similar to the way the *Space Shuttle* is currently lifted off. *Bachem* and *Fiedler* were students/friends in aeronautics at Stuttgart. Both joined the *Gerhard Fieseler Werke* after graduation.

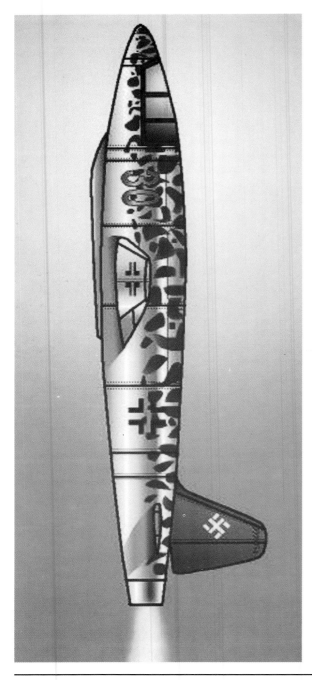

A pen and ink illustration of second version of the *Bachem/Fiedler Fi 166* known as *Design II*. It was to have been launched by its own power via a metal frame launching ramp similar to the one used to launch the "*Natter*" in late 1944 early 1945.

Erich Bachem, when he was the Technical Director of the Gerhard *Fieseler Werke*, said that he watched the Allied bombers pass overhead on numerous occasions and sought ideas on how this menace could be stopped. His "*Natter*" was an updated version of his earlier *Fi 166 II* vertical take-off manned missile. He proposed his "*Natter*" to the *RLM* to help stop American *B-17s* from dropping their destructive bomb loads from one end of Germany to the other.

A close-up view of three *Boeing B-17s* doing exactly what *Erich Bachem* and others had feared. *Bachem* believed that the only way to get past the *B-17*'s long range fighter escorts and stop the bombers dead in their flight path was through a vertically launched manned interceptor. But each time he submitted a proposal based on *Werner von Braun's* models of June 1939 and May 1941, the *RLM* rejected them.

18   The Bachem-Werke Ba 349 "Natter"

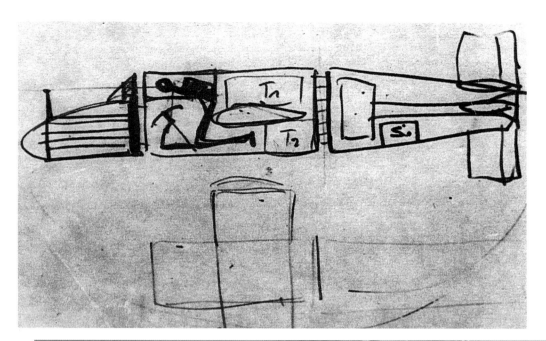

Bachem's hand-drawn "*Natter*" idea code-named *BP-20 (Ba 349V)* of mid-1944 for a vertically launched Allied bomber interceptor. Notice the similarity between *Bachem's BP-20* sketch and that of his earlier *Fi 166* proposal and *Dr.-Ing. von Braun's* interceptor from May 1941.

A port side view of the *Ba 349B* or *Mark II* sitting on a wooden cradle. It's camouflage consists of *Light Blue 76* all over its upper surfaces with a dense mottle of *Gray-Violet 75*. Under surfaces were *White 21*. Scale model by *Jamie Davies*.

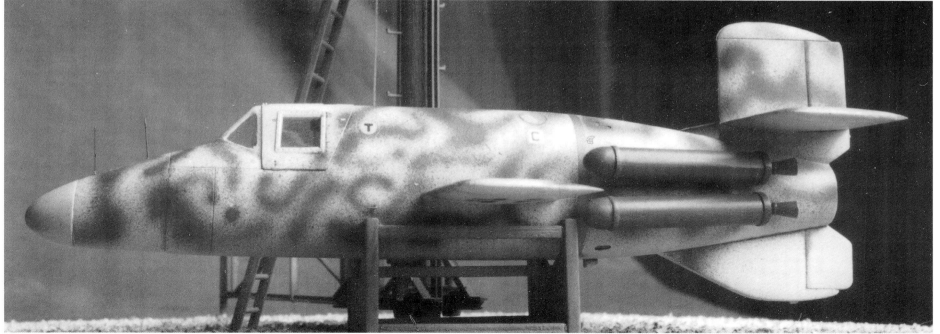

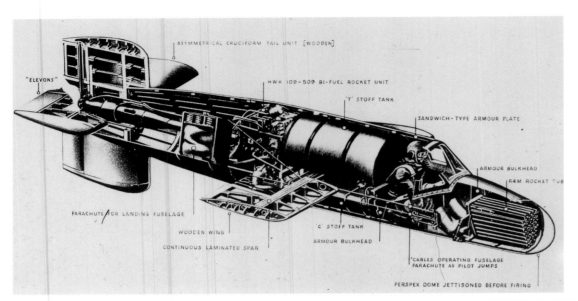

A cut-away view of the starboard side of a typical *BP-20/Ba 349V* outlining its major internal components. Left to right:
- elevators
- asymmetrical cruciform tail unit of wood
- parachute for landing the fuselage
- *HWK 509* bi-fuel liquid rocket engine
- wooden wing
- continuous laminated wooden main spar
- *T-Stoff* tank
- *C-Stoff* tank
- armored pilot's bulkhead between him and the fuel tanks
- sandwich-type armor plate protection for the pilot
- cables operate fuselage parachute when the pilot jumps out
- armored pilot's bulkhead between him and the 73 mm *R4M* rocket launcher in the nose
- *R4M* rocket shell tubes
- plexiglass dome over the *R4M* rocket shells which was jettisoned before firing

Willy Fiedler is seen fifth from the left in this 1940 photo of *Fieseler Flugzeugwerke* test pilots. L/R: *Hlenwitz, Schwalbe, Biedermann, Gerhard Fieseler,* chief test pilot *Willy Fiedler, Riedig, Wallischeck, Genthner,* and *Seidemann.*

Dipl.-Ing. Henry Bethpeder. He was the project head of the *Bachem Werke* and was in charge of the *Ba 349*'s design development. Most of the information written about the *Ba 349* program post war came from *Bethpeder*'s interrogation by *General George McDonald*'s intelligence-gathering personnel. In addition, a lot of the current myths surrounding the "*Natter*" came from several articles featuring an interview of *Bethpeder* and published in Dutch newspapers as "a hero or fanatic." He claimed to have been an unwilling participant in the "*Natter*" program and in fact said he had been sent to a *KZ* due to his sabotage of the project. *Flessner*, *Zübert*, and *Fiedler* see *Bethpeder* differently in that they do not recall *Bethpeder* ever openly opposing the war or being sent to a *KZ* for attempting to sabotage the "*Natter*" project.

Right: *Erich Bachem* envisioned a simple 70 foot tall pine tree stripped of its bark as the *BP-20*'s launcher once serial production got underway at small shops all over Germany. Pine trees were to be found all over Germany. His "*Natters*" could be transported to where they were needed...a 70 foot tall pine tree could be cut down, stripped of its bark and made into a launcher for a *BP-20*.

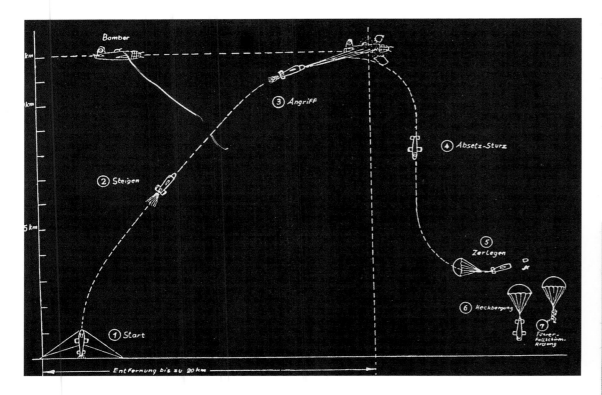

Normally the *Ba 349* would lift-off at the time the Allied bomber formation was nearest to its starting point. Upon lift-off it would be flown in an inclined path towards the bomber. After the formation had been attacked with its 24 *73* mm *R4M* rocket shells at a range of [50 to 100 meters], the *Ba 349* would go into a dive guided down to about [3,000 meters] if possible over suitable country side. Since this aircraft could not be flown as a glider due to its extremely high sink rate once the pilot fired its *R4M* rocket shells he needed to make immediate plans to abandon the machine. The next sequence would be the jettisoning of the aircraft/missile's nose which then fell away. The pilot released his safety harness and folded the control stick/column forward. With this action the "*Natter's*" rear engine/tail section would break off and fall away. The pilot would be automatically pulled out of his cockpit, too, and once clear of the aircraft, he would then open his own parachute and land in the normal manner.

A poor quality photo of *Ernst Heinkel AG's Dipl.-Ing. Karl Butter* (right) with the *RLM's* chief of Technical Development *Ernst Udet*. *Karl Butter* was the inventor of the explosive bolt. About 1941.

Karl Butter. Munich, West Germany, 1984. Photo by the author.

An artist's pen and ink illustration of a typical *Ba 349* firing off its 24 *R4M* rocket shells in one salvo into an Allied bomber formation. Each one of the *73 mm R4M* rocket shells carried 4 grams of the high explosive powder Hexagen.

If the "Natter's" R4M 73 mm rocket shells missed their mark the pilot was instructed to ram a B-17 as depicted in this pen and ink illustration, however, bailing out moments before impact. *Willy Fiedler* stated during interrogation post war that at war's end as many as 200 members of the *SS* had volunteered to pilot the *Ba 349* and were willing to give their lives if necessary to stop the destruction being caused by aerial bombing.

This is what *Erich Bachem* and *Reichsführer-SS Himmler* believed would be the result after a "*Natter*" had released its 24 *R4M* rocket shells. Here we see a *B-17* with its starboard wing blown off at its wing root by a *Flak* shell.

This *B-17* has also suffered a direct hit from a *Flak* shell—the same results *Bachem* and *Himmler* believed would happen from a direct hit by a single *R4M* rocket shell—and has rolled over on its back as its entire port wing is falling away.

Bachem's initial flight launching/testing of the *Ba 349* was done with a permanent metal frame vertical ramp as shown in this photograph. Later, *Bachem* wanted to use a single 70 foot tall pine pole to launch his "*Natter*" from sites all over Germany. In early 1945, the *RLM* had ordered 1,000 cockpit instruments per month to be delivered to *Ba 349* manufacturing facilities. So, it appeared, that *Bachem* and the *SS* believed that they had a workable and effective machine for countering the increasing number of Allied bomber formations over Germany.

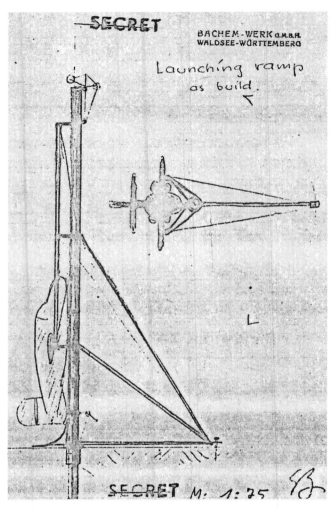

A poor quality pen and ink illustration of *Erich Bachem's* simple fresh-cut pine pole launching ramp which could be set up quickly almost anywhere inside Germany for the rapid launching of the "*Natter*" once they became fully operational.

A successful launch of *Erich Bachem's Ba 349A* manned missile with *Lothar Sieber* in the cockpit. The twin trails of dark smoke come from the 4x*Schmidding 533* solid-fuel rocket boosters. The *HWK 509* is leaving a single trail of white smoke between the *Schmidding* booster rockets.

The headquarters of the *RLM*-Berlin where *Bachem* took his proposal for the *BP-20*. This photograph was taken shortly after the building was completed in the late 1930s.

This is how the *RLM* building (center of picture) looked post war after the Red Army's shelling of Berlin. The street has been cleared of rubble so military vehicles can make their way without obstruction.

Above: Three other unofficial boss' of the *RLM*: *Roluf Lucht, Gottfried Reidenbach,* and *Helmutt Schelp*. All in all there were about six leaders within the *RLM* who labeled themselves as members of "the club," and as *Helmutt Schelp* once told this author in a telephone interview, "we ran the show, and that show didn't include *Erich Bachem's* vertically launched manned missile...at least not up until *Reichsführer-SS Himmler* intervened."

Left: *Oberst Siegfried Knemeyer*, the very powerful, highly intelligent, and influential director of *RLM's* Technical Office and the *RLM* in general. In 1944/1945 he was pretty much the "godfather" of the *RLM* and from this photo of *Knemeyer*, his sly smile represents the general reaction *Erich Bachem* received when he presented them his idea of the *BP-20*...bemusement bordering on laughter.

Erich Bachem sought the support of *General der Jadgfliegers Adolf Galland* when "the club" turned him down flat. Although *Galland* is reported to have been interested in seeing the machine receive *RLM* support he was not one of the decision-making insiders at the *RLM* at that time, and *Bachem's BP-20* rejection remained unchanged.

Dr.-Med. Siegfried Ruff, director of flight medicine for the *RLM*, said that a human pilot could withstand 2.2 times gravity expected during take-off of the "*Natter*" after leaving its launch pad. In addition, the pilot would be laying on his back throughout take-off allowing him to take even higher "g" forces before passing out. Consequently, the pilot's position in the vertically launched rocket-powered *BP-20* interceptor was very good. *Dr. Ruff*, a sailplane enthusiast from Bonn, had been a life-long friend of *Reimar Horten* of the *Horten* brothers and their all-wing sailplanes, also from Bonn.

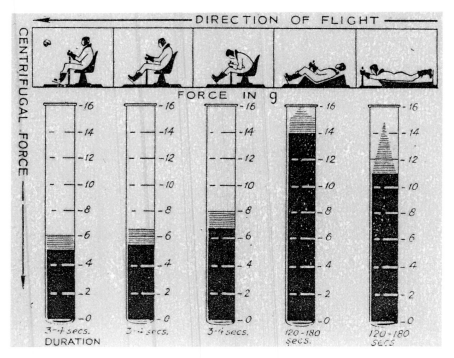

Graphically shown in this series of photos are a pilot's reaction as the amount of gravity he experiences in various seating positions. *Dr. Ruff* stated that at 2.2 times gravity experienced by a "*Natter*" pilot after lift-off, the pilot would not likely "black out" nor lose control of his machine. The worse position to withstand high "g" is in an upright sitting. The superior position is where the pilot is able to lay on his back. *Dr. Ruff*, through his extensive research found that a pilot on his back could withstand up to 14 "g's" for as long as 120 to 180 seconds without blacking out. What happen to *Oberleutnant Sieber* during the first manned test flight of the *Ba 349 M-23* on 18 February 1945 remains a mystery to this day. However, it appears that *Sieber's* cockpit canopy with the attached head rest flew off. In doing so *Sieber* may have suffered his head injury's rendering him unconscious.

Erich Bachem, undaunted after *Galland* failed to sway the *RLM's* club, sought out *Reichsführer-SS Heinrich Himmler* and received an audience with perhaps the single most feared individual in the *Third Reich* and presented his plans for a vertically launched interceptor. *Himmler* shown here in this photograph is second from the left. Next to *Himmler* is another highly feared *SS* individual...*Reinhard Heydrich* the so-called "Butcher of Prague."

30   The Bachem-Werke Ba 349 "Natter"

Bachem described to *Himmler* how Allied bomber formations such as American *B-17s* and not their fighter escorts, would be the target of his proposed *BP-20* rocket-powered interceptor. The *P-47s* and *P-51s*, said *Bachem*, would be unable to shoot down the "*Natter*" as it made its very high-speed approach into the *B-17* bomber pack.

Bachem showed *Himmler* illustrations how the *BP-20* would lift-off supported by a freshly cut pine tree placed in the ground at locations anywhere within the *Third Reich* where Allied bombers were concentrating their formations.

A rack of the non-guided powder rocket, the *R4M* seen here mounted under the wing of a *Me* 262. The *R4M* was invented by *Dipl-Ing. W. Perniss* of the *Deutsche Waffen-und Munitionsfabrik*, Lübeck. It had a caliber of *73* mm and radially arranged self-stabilizing fins. It could achieve a forward speed of 250 meters/second and was typically fired from ranges of between 1,500 and 1,800 meters. It weighed only 8.75 pounds [4 kilograms] including its Hexogen-charge (tetramethylene-trinitramine) of 500 grams. This rocket shell was detonated by a proximity fuse fitted in its nose. Its success exceeded all expectations.

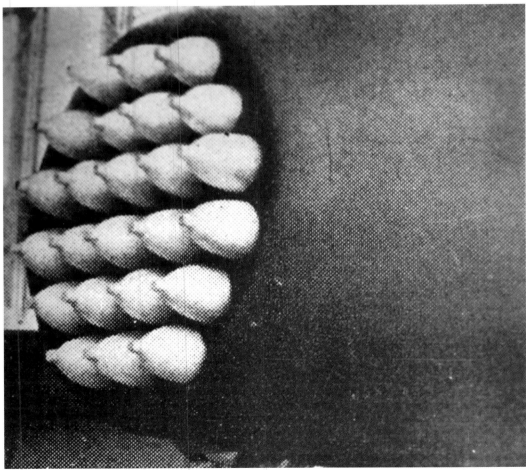

A port side nose view of the twenty four *R4Ms* arranged in the nose of the *Ba 349A*. They were covered during lift-off by a jettisonable plexiglass nose-cone and were all fired simultaneously by electricity and detonated by a proximity fuse.

A direct on nose view of the *Ba 349A* with its twenty four *R4M* rocket shells as seen at Freeman Field post war. The battery of rockets were aimed by a crude ring sight which can be seen a few feet aft the *Ba 349's* nose.

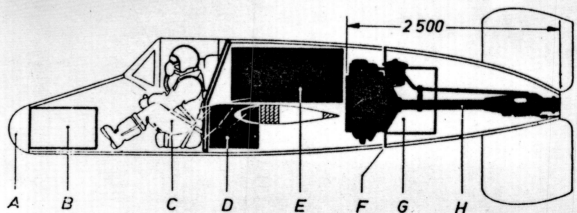

A pen and ink drawing of the *Ba 349A* as seen from its port side.
a. plexiglass nose cone
b. rocket tubes containing the 24 *73* mm *R4M* rocket shells
c. pilot and seat position
d. *C-Stoff* tank
e. *T-Stoff* tank
f. HWK 509's propellant waste drain after shut down
g. HWK 509's rocket motor housing
h. HWK 509's thrust tube connecting the rocket motor and the combustion chamber

Einbauschema der Walter HWK-109-509. A) Abwerfbare Plexiglashaube, B) Raketenbatterie, C) Pilot, D) „C-Stoff"-Behälter, E) „T-Stoff"-Behälter, F) Sollbruchstelle des Rumpfes (absprengbar), G) Fallschirmkasten (Steuerbord), H) Triebwerk HWK 109-509 A. (Zeichnung Kens)

A fully assembled *HWK 509* bi-fuel liquid rocket engine on a test stand. The entire engine weighed only 374 pounds [170 kilograms].

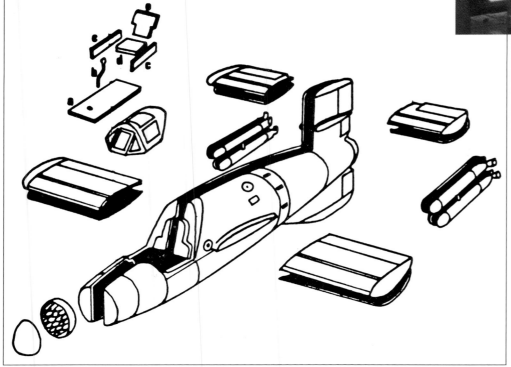

A pen and ink illustration of the several wooden component parts for a complete *Ba 349A* "*Natter*" and how they all fit together during assembly.

Erich Bachem (left) and *Waffen-SS General Wolff* (center). *General Wolff* was general director of the *Ba 349* project.

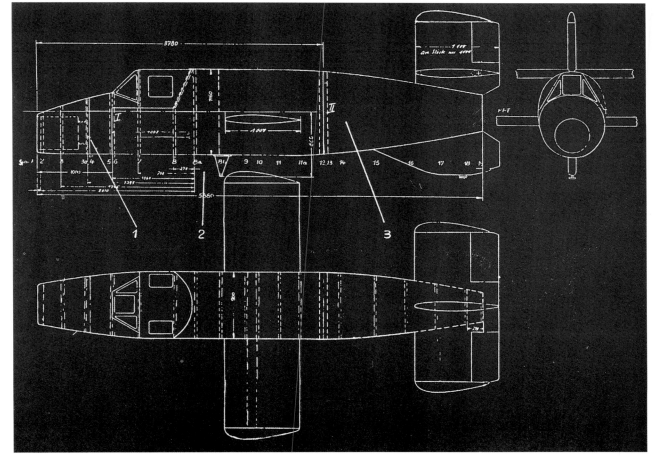

A pen and ink three-view exterior drawing of the *Ba 349A*.

The Bachem-Werke Ba 349 "Natter"

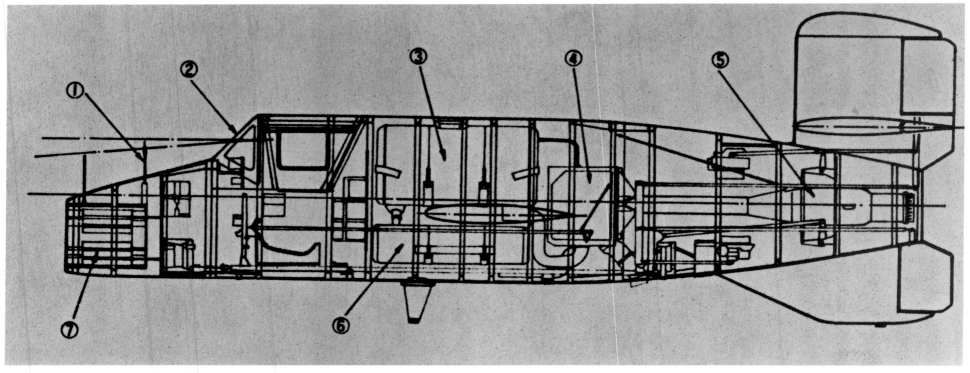

A pen and ink port side view interior drawing of the *Ba 349A*.
1. ring sight
2. armored cockpit windscreen
3. *T-Stoff* tank
4. *HWK 509* bi-fuel liquid rocket motor housing
5. *HWK 509* combustion chamber
6. *C-Stoff* tank
7. rocket tubes for 24 *73* mm *R4M* rocket shells

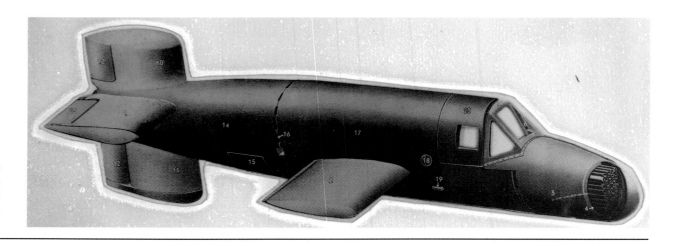

The starboard side fuselage, wing, and tail assembly of a *Ba 349V* from *Erich Bachem's* failed proposal to the *RLM*. *Bachem* illustrated his proposal with the stunting *Hans Jordanoff* dissected views. *Jordanoff* was an employee of the *Bachem Werke*.

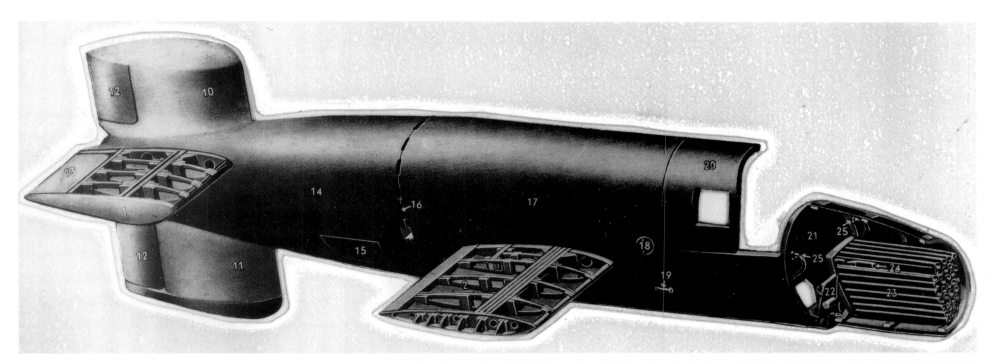

The *Jordanoff*-drawn starboard side of a *Ba 349V* minis the cockpit windscreen but featuring internal views of the horizontal stabilizer, wing, and in the nose, its armored bulkhead, instrument and rocket collector, and the *73* mm *R4M* rocket tubes.

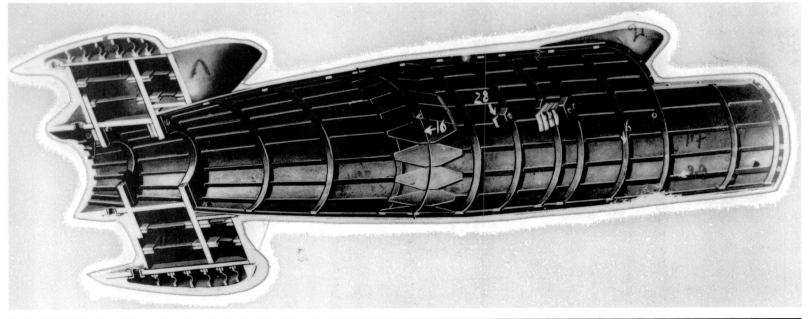

The port side fuselage of a *Ba 349V* in *Jordanoff* featuring its ribs, stringers, and skinning. The diamond-shaped items (*#16*) surrounding the fuselage are the fuselage separation joints (break points) between the tail assembly and the wing area of the fuselage.

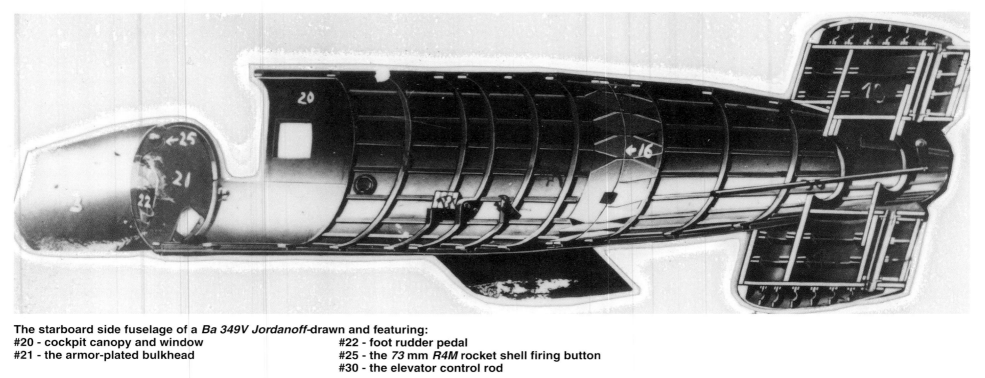

The starboard side fuselage of a *Ba 349V Jordanoff*-drawn and featuring:
#20 - cockpit canopy and window
#21 - the armor-plated bulkhead
#22 - foot rudder pedal
#25 - the *73 mm R4M* rocket shell firing button
#30 - the elevator control rod

The starboard side of a *Ba 349V Jordanoff*-drawn and featuring:
#32 - fuselage parachute container,
#38 - *T-Stoff* tank
#43 - rear bulkhead
#45 - pilot in a protective suit

Brigadier-General George McDonald USAAF's Director of Intelligence shown leaning on the starboard horizontal stabilizer of a *Fw 190D* formerly of *JG4* which had been recovered near Frankfurt post-war.

L/R: *SS General Frank, Reichsführer-SS Himmler*, and *Waffen-SS General Wolff*. General Wolff was in charge of the *Ba 349* program, its development, serial production, testing, and operational readiness all taking place at the *Bachem Werke*, Waldsee/Württemberg.

Waffen-SS Oberleutant Dipl.-Ing. Heinz Flessner. The 33 year old was in charge of 200 "wounded" *Waffen-SS* personnel assigned to the *Ba 349* program at *Bachem Werke,* Waldsee/Württemberg. *Flessner* told this author that the job of the *Waffen-SS* at the *Bachem Werke* was find food, supplies, fuel, tools, trucks, and so on needed to support the "*Natter*" project. A very difficult job at that late hour of the war. *Flessner* and his men would fan out to neighboring towns, farms, and villages to physically obtain—with force, if necessary—from anyone anywhere all the items the *Bachem Werke* required to develop and test the "*Natter*." "This was the job of the *Waffen-SS* men assigned to the *Bachem Werke.*

Waffen-SS Oberleutant Heinz Flessner (center) leading his *SS* work crew before individual groups fan out to the neighboring country-side scavenging for supplies needed by the *Bachem Werke.* Winter 1944.

Right: *Waffen-SS Oberleutant Heinz Flessner*. Although it was late 1944/early 1945, the *Waffen-SS* personnel at *Bachem Werke* still carried on their strict dress code and personal appearance as this picture shows. *Flessner* had served with the *Waffen-SS* at the Russian Front, was wounded, and sent back to Germany for hospitalization. Had he not been wounded, he told this author, in all likelihood he would have been captured by the Soviets as were his entire division and later died in captivity as they did. In post war Germany all living members of the former *SS* are socially and economically penalized/condemned for the sins of their *Reichsführer-SS Heinrich Himmler,* he told me, although the *Waffen-SS* and the *KZ-SS* guards running the numerous *KZs* throughout Europe were entirely separate...the *Waffen-SS* being formed in September 1939. He joined the *Waffen-SS* in November 1939.

Below: The entire 200-man *Waffen-SS* contingent assigned to the *Bachem-Werke* assembled for morning roll-call.

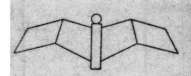

The personal identity card #41051 carried by *Waffen-SS Oberleutant Heinz Flessner* while he was assigned to the *Bachem Werke*. His card had been issued on 21 October 1944. He remained with the *Bachem* group until he and a dozen others surrendered near Salzburg, Austria to the American Army's 44th Division along with three or four brand new "*Natters*."

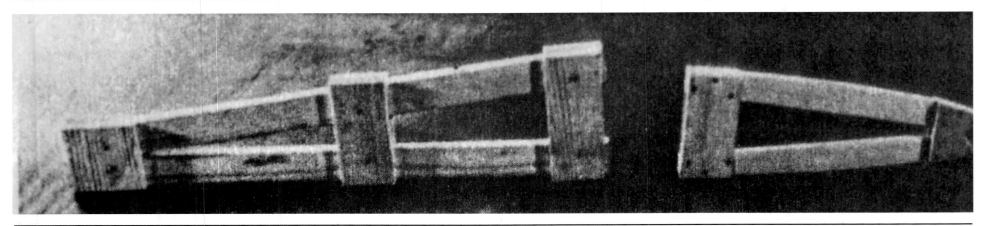

The very basics of the *349's* internal wing...crude, rough cut wing ribs.

The *349's* 4xboard sandwiched wooden main spar seen in the upper right-hand corner. The wing ribs were glued on to the main spar's leading and trailing end.

The starboard side wing providing a good view of its 4xboard sandwich main spar. It appears that the internal structure is complete except for a piece of wood which will fit into each rib's leading edge which is seen up at the top of the photograph.

The Bachem-Werke Ba 349 "Natter"

Several of the individual rib blocks used in the construction of the "*Natter's*" wing tips.

A *Bachem Werke* laborer is working on what appears to be the port wing tip for a *Ba 349* at Waldsee/Württemberg.

What appears to be a complete wing, except for covering, for a *Ba 349* clamped to its assembly table at the *Bachem Werke*, Waldsee/Württemberg.

A molded one-piece plywood covering which wraps around the leading edge of a "*Natter's*" wing. Shown in the *Ba 349* workshop at the *Bachem Werke*, Waldsee/Württemberg.

A complete wooden wing for a *Ba 349*. It appears that its wing tips have yet to be glued on. Both wings are connected by 4xboard sandwiched main spar. The *349's* wings carried no ailerons. *Bachem Werke*, Waldsee/Württemberg.

Fore and aft ribs used in construction of the "*Natter's*" horizontal stabilizer.

A *Bachem* laborer working on a wing for the *Ba 349* at Waldsee/Württemberg. He appears to be using a yard long sanding board to make sure that all the leading edge ribs are of uniform height after which it will be covered by a thin sheet of plywood.

Left: A complete, but as of yet, an unpainted horizontal stabilizer for an early version of the *Ba 349V.*

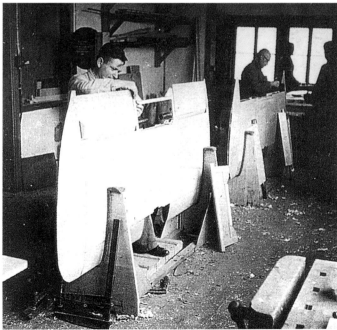

Right: Several workers appear to be putting final touches to the horizontal stabilizer for a *Ba 349* at the *Bachem Werke,* Waldsee/Württemberg. A second horizontal stabilizer appears in the background.

The sequence (a bit out of place, however) in which a "*Natter's*" horizontal stabilizer's elevators were constructed: ribs, completed elevator, uncovered elevator, and horizontal stabilizer end cap/tip.

Left: A *Bachem Werke* laborer, perhaps holding a glue pot, is putting final touches to a "*Natter's*" horizontal stabilizer. The metal pipe which connects each elevator can be seen to the right of the laborer's hand.

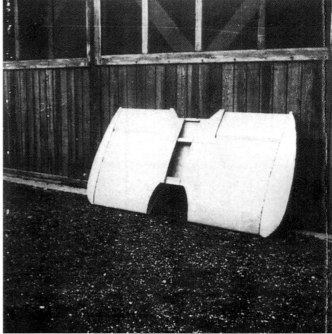

Right: A rare photo showing the vertical stabilizer of an early *Ba 349V* model. Both upper and lower portions were built as one piece and cut in half prior to installation on the aft fuselage. The "*Natter's*" vertical stabilizer was substantially changed beginning with the *M-16*.

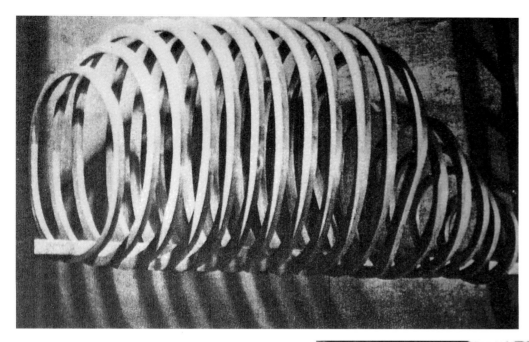

A complete set of wooden bulkheads (the nose bulkheads for a Ba 349 begin to the right in the photo) around which to construct the fuselage for a "*Natter*."

This is how the fuselage for a *Ba 349* was built up...each on its own individual construction jig.

The aft fuselage featuring its bulkheads and stringers of an early model "*Natter*" at the *Bachem Werke*, Waldsee/Württemberg.

A poor quality photograph of the starboard side nose of a *Ba 349's* fuselage under construction. Shown are the numerous uncovered fuselage ribs but pretty much minus its stringers. The wings have been installed, however, perhaps for fit and alignment. *Bachem Werke*, Waldsee/Württemberg.

A starboard nose view of a "*Natter*" under construction, however, minus its *R4M* hexagonal rocket tube package and nose cone. Inside the fuselage can be seen the pilot's control stick.

A nose port side view of a *Ba 349's* uncovered fuselage and nose area. In addition to the wings being installed, it appears that the tail assembly is attached, too. *Bachem Werke*, Waldsee/Württemberg.

The armored windscreen of the "*Natter*" and it appears to have been recently installed. It fit around and over the wooden windscreen framework.

The pilot's control stick shown in a *Ba 349* under construction at *Bachem Werke*, Waldsee/Württemberg.

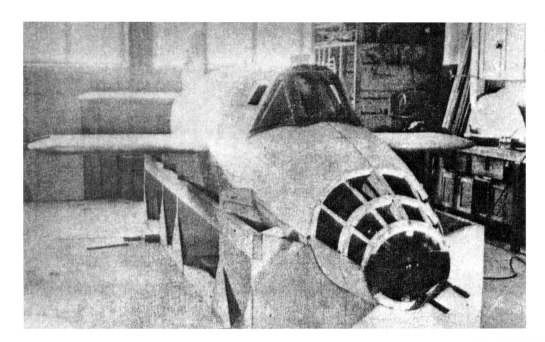

A poor quality photo of a *Ba 349* as seen from its starboard nose side. The *R4M's* hexagonal rocket tube container appears to have been installed and waits for its plywood skin and nose cone.

A *Ba 349* fuselage being attended to by as many as four female workers at the *Bachem Werke*, Waldsee/Württemberg. *Erich Bachem* is shown in the white lab coat. Each of the people in the photo have on heavy clothes so temperature in the work shop must have been cool. Notice, too, a completed cockpit canopy up on the shelf seen in the far right in the photo. About 600 people worked at *Bachem Werke* including 200 members of the *Waffen-SS*.

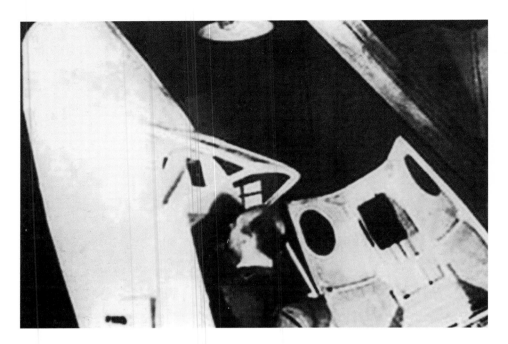

One of the *Bachem Werke* laborers is shown inside the unfinished cockpit cabin of a *Ba 349*. This "*Natter*" appears in a curious position as if it were in a nose-up attitude probably because it is mounted on a rotateable work stand.

The starboard side of a *Ba 349* under construction at the *Bachem Werke* featuring its tail cone assembly. This assembly could easily be attached or removed from the fuselage so that maintenance could be done on its *HWK 509* bi-fuel liquid rocket-engine.

What appears to be recently constructed tail assemblies for the "*Natter*" at *Bachem Werke*, Waldsee/Württemberg. From the photo it is not possible to identify them as being *349As* or *349Bs*. The tail assembly in the center is numbered "*8*" while the one directly behind is #*9*. Notice that the item in the bottom center appears to be a frame assembly for the cockpit windscreen sitting on its nose.

A poor quality photo of a wooden tail assembly nearing completion for a *Ba 349* at the *Bachem Werke*, Waldsee/Württemberg.

A close-up view of the tail assembly for a "*Natter*." Notice how the thin plywood has been wrapped around the fuselage bulkheads after being secured with wood glue and nails.

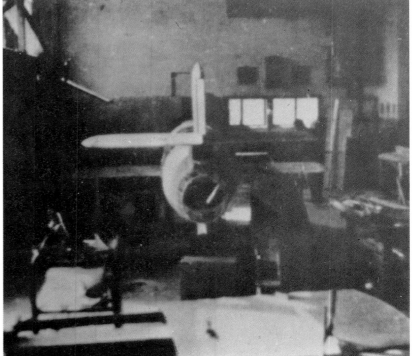

A poor quality photo of a new *Ba 349A* nearing completion at the *Bachem Werke*, Waldsee/Württemberg and featuring the tail assembly minis its tail cone.

An overhead view in the *Bachem Werke*, Waldsee/Württemberg where several *Ba 349's* appear to be in their final stage of completion. Notice that none of the *349's* shown carried a *Balkenkruez* or *Halkenkreuz*. The *349* in the center of the photo has its tail assembly removed and showing its installed *HWK 509* bi-fuel liquid rocket engine.

A poor quality close-up view of the *Ba 349A* shown in the above photo featuring the area around the *Ba 349A's HWK 509* combustion chamber which appears to be requiring plywood skinning.

A very poor quality photo of the work shop at the *Bachem Werke*, Waldsee/Württemberg and featuring several *349s* appearing ready to be shipped to their launch pad.

A very poor quality photo of two *Ba 349's* under going completion near the wall in the darkened *Bachem Werke*.

A poor quality photo of the *Bachem Werke* shop at Waldsee/Württemberg with several *Ba 349s* apparent in the fore ground.

A poor quality photo of what appears to be several fully assembled *349As* amid all around junk. This author does not know the location of these "*Natters*." Perhaps it was the *Bachem Werke* at Waldsee/Württemberg.

DVL-Adlersholf (Berlin) readily agreed to test the *BP-20* in their wind tunnel...and who wouldn't knowing that the *Reichsführer-SS* himself was now enthusiastically involved? Shown in this photo is the main entrance to *DVL*-Adlersholf.

The mouth of the giant wind tunnel at *DVL*-Adlersholf where *Erich Bachem* took his 1/4 scale model of his *BP-20* for wind tunnel testing.

A nose-on view of the "*Natter*" wooden model used in the *LFA's* high-speed wind tunnel.

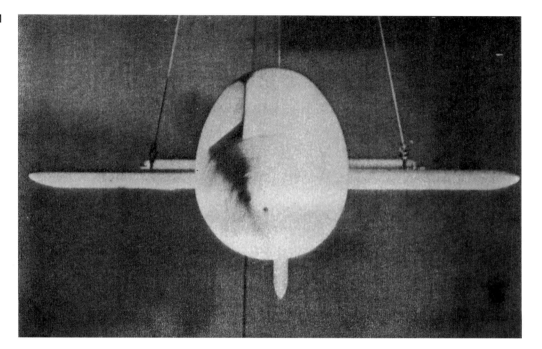

Erich Bachem was also given the use of the high-speed wind tunnel at *LFA*-Brunswick. Shown in this photo is *Bachem's* 1/4 scale model *BP-20/Ba 349V* in the 9.1 foot [2.8 meter] diameter high-speed wind tunnel at *LFA*-Brunswick during March 1945. Wind tunnel tests were conducted on the complete model, wing alone, and tail section alone with various elevator settings.

A copy of *LFA's* 3-view drawing regarding their wing-tunnel testing of the *Bachem BP*-20's tail assembly and the results expected were the vertical and horizontal surfaces enlarged by degrees from its design specifications.

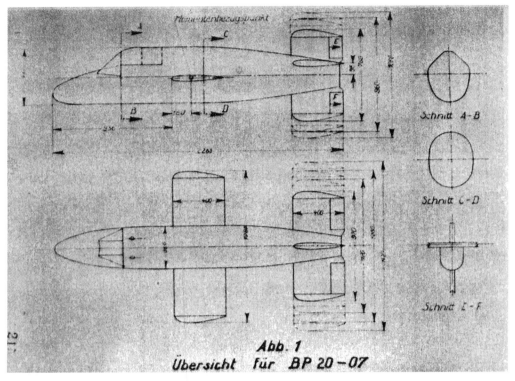

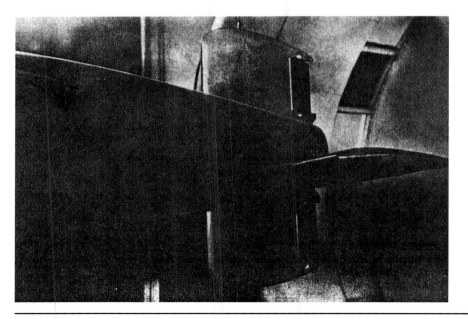

A port side view of the 1/4 scale model of the *BP-20's* tail assembly as tested by *LFA*, March 1945.

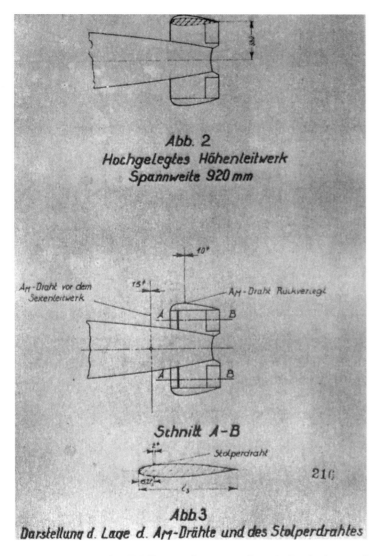

Above: A copy of *LFA*'s illustration regarding their wind-tunnel testing of the *Bachem BP-20*'s tail assembly.

Right: A series of photos involving an early *349V* of the type used in *Bachem's* initial unmanned flight test program. This *349V* has been mounted in the metal launching ramp featuring its starboard side at Heuberg prior to an unmanned test flight by 4x*Schmidding 533* booster rockets. Notice that the cockpit has no windows but plywood instead. This author is unable to identify the two civilians shown in the photograph.

A starboard side view close up of the *349V* seen above. This unmanned machine would have been powered only by its 4x *Schmidding 533* booster rockets. It appears to being inspected by a small group of men.

Another *349V* as seen above but with a close-up of its tail cone where the *HWK 509* would extend out. As many as four *Bachem-Werke* laborers plus a man in a light-colored hat appear to be concerned with some aspect of the middle launch rail where the *349V's* ventral tail rides during lift- off.

Above: An unmanned *349V* being made ready for lift-off by only 4x *Schmidding 533* solid fuel booster engines. The "*Natter*" carried no *Balkenkruez* or *Halkenkreuz* during *Bachem's* initial flight testing program or after.

Right: A *349V* "*Natter*" in its metal frame launcher with everything appear black against a gray background.

Above: Lift-off of an unmanned *349* during *Bachem's* initial flight testing program. At the beginning several unmanned "*Natters*" failed even to clear the metal frame launch tower. The reason was due to the differential thrust provided the 4x*Schmidding 533* booster rocket engines. After much trial and error the booster rockets became better coordinated.

Right: A *349V* unmanned flight test machine known as *M-17*. It has the same words stenciled on each of its elevators: "If found, report to Commander of Troop Exercise Control, Heuberg. Telephone: Stetten am Kalten Markt 222. REWARD!"

The Ba 349 M-17 has cleared its launch tower.

A close up of the unmanned M-17 in vertical flight. Its 4x Schmidding boosters would operate for approximately 12 seconds and then fall off thanks to explosive bolts developed by Karl Butter of Ernst Heinkel AG.

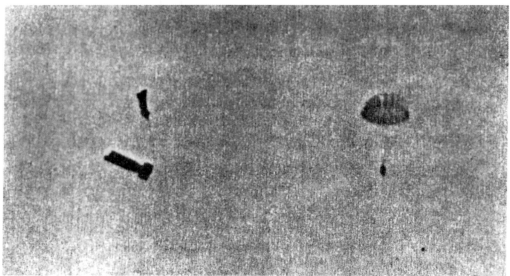

Hans Zübert told this author that he made four unpowered glider flights in *Ba 349Vs*. The plan for one test flight was to have him bail out of the cockpit and descend to the ground via parachute. The "*Natter*" was also lowered to the ground via its own parachute. This is a photo of *Hans Zübert* (right) and the "*Natter*" (left) with its parachute about to open.

An unmanned 349V, with numerals *2+2* painted on the upper surfaces of its wing, appearing on its metal frame launch pad at Heuberg. Several *Bachem Werke* laborers are attending to some aspect of the machine prior to its take-off. Notice the unknown cylinder-shaped projection on this "*Natter's*" nose.

Hans Zübert, highly experienced sailplane pilot, on loan to *Bachem Werke* from the *Horten Flugzeugbau*-Göttingen. It is was *Zübert's* job to test the gliding characteristics of the "*Natter*" which he did between June/November 1944. *Bachem* had only two pilots to help him test fly the *Ba 349*: he and *Lothar Sieber*. Prior to joining *Bachem*, *Zübert* had been to Rechlin where he had flown the *Me 163* and *Me 262* in preparation to piloting the "*Natter*."

A pen and ink illustration featuring how *Hans Zübert* would have seen the "*Natter*," with its own parachute open, as he falls away before opening his parachute.

A port side, close-up, view of an unmanned 349V on its metal frame launch ramp. This author does not know the "M" number of this Bachem test machine. Waffen-SS Oberleutnant Flessner told this author that Erich Bachem placed dogs in the cockpit of several unmanned test "Natters."

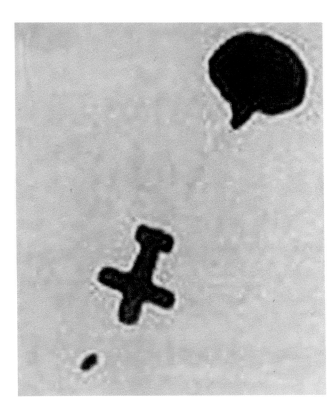

The actual situation illustrated above. *Hans Zübert* is seen falling away from a "*Natter*" whose parachute has opened fully. *Zübert's* parachute opened moments after this picture was taken.

The *349V M-8* with its fixed tricycle gear (main wheels came from a *Klemm Kl 35*) minus its *HWK 509* or *Schmidding 533* power plants. The *M-8* was used for gliding flight testing and was piloted by *Hans Zübert*. He told this author that he had piloted the *M-8A* and *M-8B* and that *Lothar Sieber* test flew *M-9* (*M-23*).

The test aircraft's parachute installed where the *HWK 509's* combustion chamber would be in the *M-8A* and *M-8B*...both piloted by *Han Zübert* to gently lower the wooden flying machine to the ground.

Here, *DFS Flugkapitän Hans Zacker* in a *He 111* has a cable attached to a dorsal hook about the center of gravity of the *M-8* being piloted by *Horten Flugzeugbau* test pilot *Hans Zübert*. This means of towing is known as "carrier towing" and it was developed by DFS. Later *Zübert* landed the *M-8* but he recalled that given the wing's extremely high wing-loading, the *M-8* had a very short glide angle because it was sinking like a rock.

Several unmanned test versions of the "*Natter*" had their parachute located where the *HWK's 509 T-Stoff* and *C-Stoff* fuel tanks would normally be.

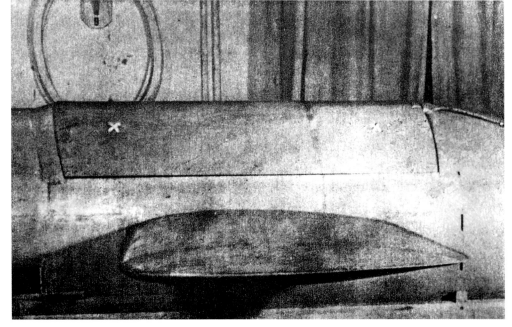

A poor quality photo of a unmanned *Ba 349V's* fuselage as seen from its port side wing tip. The wing's leading edge of this "*Natter*" is on the left. The molded one-piece upper fuselage-rib plywood covering and shown with an "*X*" on each end, has not yet been fastened down. *Bachem Werke*, Waldsee/Württemberg.

The center fuselage and tail assembly of one of the engine-less *M-8* "*Natters*" which had been piloted by *Hans Zübert*, "carrier towed" to altitude by the *DFS Heinkel He 111*, and allowed to parachute to the ground after *Zübert* bailed out. This "*Natter*" fuselage appears to be in relatively good shape after the ordeal. This author does not know how the *Bachem Werke* test dogs fared after their ordeal.

One of the four initial test *Ba 349Vs* up on a wheeled tricycle cradle to carry it about and during take-off by "carrier tow" cable from a *DFS*-flown *Heinkel He 111* tow plane. The *Ba 349V's* starboard side nose is featured.

A *Ba 349V* manned but unpowered test machine featuring its port side nose. The white rope is for pulling the *349V* around the test sight when it is on its tricycle dolly and then to line it up for its *He 111* tow aircraft.

The starboard side rear of one of the four engine-less "*Natter's*" seen mounted here on its tricycle wheeled transporter/carrier.

Three unidentified *Bachem Werke* civilians and *DFS's Flugkapitän Hans Zacker* (right) standing out front of what appears to be one of the two *M-8s* (M-8a and M-8b) which had been piloted by *Hans Zübert*. Notice, too, in the background is a *Me 262!* This photo was taken at *DFS*-Ainring and it would be interesting to know what the *Me 262* was doing there at this late hour of the war. Post war the American Army found one complete but broken up *Ba 349* at *DFS*-Ainring. It may have been one of the two *M-8s*.

A poor quality photo of the *Ba 349 M-13* on its metal frame launching tower at Waldsee/Württemberg.

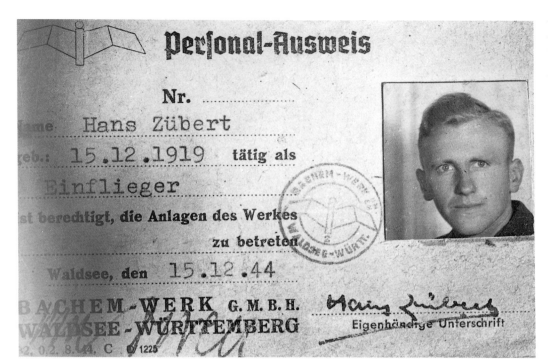

The ID card issued by the *Bachem Werke* for 25 year old *Horten* sailplane test pilot *Hans Zübert*. He told this author that *Erich Bachem* and the *SS* were making plans to start building the piloted *Fieseler Fi 103* flying bomb. In addition, *Bachem*, was redesigning his "*Natter*" into a prone-piloted position. *Zübert* said that no prone position "*Natters*" were completed prior to war's end although *Willy Fiedler* was in the process of converting one *Ba 349A* fuselage to prone piloting.

Heuberg (near Regensburg), Spring 1945. *Hans Zübert* is seen on the far right sitting with his hands on his knees. In the background is a *Ba 349A* mounted on a prototype 70 foot high pine tree launch pole.

Hans Zübert (right) is shown sitting on a tanker wagon. He told this author that he recalled only one scary flight while testing the "*Natter.*" *Zübert* had been towed into the air by the *DFS He 111* tow plane. At the end of the flight the final test was to detach the cockpit from the rest of the fuselage, however, the cockpit/fuselage locking mechanism would not release. He tried and tried while seconds passed and finally it separated. Then, as the cockpit cabin was falling to the ground with him still inside he'd release his windscreen allowing him to exit and take to his parachute. Well, the windscreen wouldn't release either, and now *Zübert* was getting scared because if too many more seconds passed then he would not have enough time to get his parachute out and deployed. However, he was able to release the windscreen with barely enough time to deploy his parachute before hitting the ground. *Zübert* told this author that if only a few more seconds had passed by he'd not be here today talking about the "*Natter*" project with me.

Hans Zübert (left) and his old friend from the *Bachem* "*Natter*" days *Heinz Flessner*. Bad Homburg, West Germany August 1987. Photo by the author.

Creator of the famed *HWK 509* bi-fuel liquid rocket engine used by the *Me 163, Me 263, DFS 228,* and the *Ba 349*...Hellmuth Walter.

Right: The pumping unit for the *HWK 509* bi-fuel liquid rocket engine. From the main fuel tanks, which were housed in the "*Natter*" behind the pilot, *T-Stoff* and *C-Stoff* entered the pumping unit. These fuels were then pressurized by the centrifugal impellers in the turbine pump group. The large diameter tube seen at right is the thrust tube. The small diameter tubes are (top) *C-Stoff* coolant outlet pipe and (bottom) *C-Stoff* coolant inlet pipe.

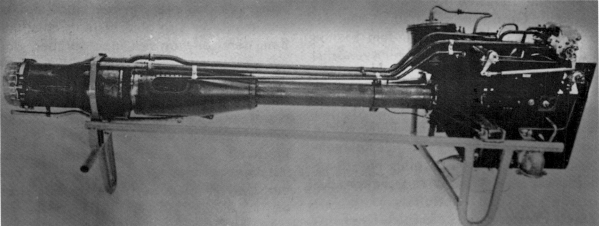

The standard *HWK 109-509A* bi-fuel rocket engine of 3,750 pounds [1,700 kilograms] thrust.

A port side view of the *HWK 509's* combustion chamber (right) and the thrust tube (left) connecting the "mixing motor" and the combustion chamber. Inside are its feed pipes to the combustion chamber. The top small diameter pipe is *C-Stoff* coolant outlet while the lower pipe is *C-Stoff* coolant return.

A good view of the exhaust rudders installed in the jet stream aft the *HWK 509's* combustion chamber orifice. A wooden tail cone would complete the installation. Either side of the of the exhaust rudders can be seen metal rods going up to the elevators so when the pilot moved the control stick after clearing the launch ramp during lift-off, he'd have a small amount of control in maintaining his flight path. The small diameter pipe to the right of the combustion chamber is a scavenge pipe used to jettison unused *C-Stoff* remaining in the combustion chamber after the *HWK 509* has been switched off.

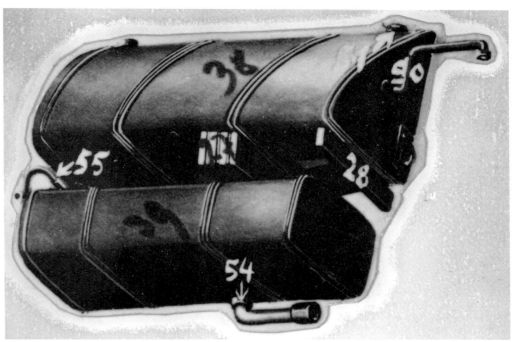

The twin separated fuel tanks for the *HWK 509* bi-fuel liquid rocket engine in a Hans Jordanoff dissected view:
- #28 - wing main spar
- #38 - 120 gallon *T-Stoff* tank
- #39 - 66 gallon *C-Stoff* tank
- #54 - outlet for the *C-Stoff* tank going to the *HWK 509* engine
- #55 - *C-Stoff* overflow pipe

From this small diameter opening came 3,750 pounds [1,700 kilograms] of neck-snapping thrust.

A port side view of the pilot, fuel tanks, and parachute container in a *Jordanoff* dissected view:
- #20 - cockpit canopy
- #42 - *C-Stoff* overflow pipe
- #50 - *T-Stoff* tank

Below: The *Ba 349V* in a *Hans Jordanoff* dissected view. Left to right:
- #32 - fuselage parachute container
- #33 - parachute opening cables
- #34 - parachute ejection springs
- #35 - parachute container catch
- #28 - wing auxiliary spar
- #42 - *C-Stoff* overflow pipe
- #39 - *C-Stoff* tank
- #27 - wing main spar
- #38 - *T-Stoff* tank
- #36 - parachute container catch release cables
- #40 - *C-Stoff* tank filler opening
- #41 - ventilation pipe
- #43 - armor bulkhead behind the pilot
- #45 - pilot in a protective suit
- #47 - pilot's shoulder strap
- #46 - pilot's waist strap
- #44 - pilot's lower seat cushion
- #60 - pilot's oxygen-air supply
- #20 - cockpit canopy
- #48 - pilot's control stick

A *Bachem Werke HWK 509* specialist is refueling a "*Natter's*" *C-Stoff* tank. This tank held approximately 66 gallons [270 liters]. Notice the protective outer clothing the man is wearing. This *Ba 349* is mounted on a 70 foot pine tree pole. The refueling filler for the *T-Stoff* tank can be seen above the two access ports to the *HWK 509* mixing motor.

A *Ba 349A*, mounted on a 70 foot pine tree pole has been refueled. Notice how the "*Natter*" was lifted up to its mounting position by a chock wrapped around each wing at its wing root. A cable extends down to each chock from a spreader bar seen here at the nose of the wooden machine. Notice, too, the front ends of the "*Natter's*" *Schmidding 533* booster rockets.

The port side of a *Ba 349A* mounted on its metal frame launch ramp. Notice the paired *Schmidding 533* booster rockets in black camouflage. Scale model by *Reinhard Roeser*.

Full scale model of a *Ba 349A* without its outer plywood paneling. This is its starboard side and it features the two starboard side *Schmidding 533* booster rocket engines without their field service end caps.

A pen and ink drawing of the *Schmidding 533* solid-fuel rocket booster engines of the type and arrangement used on the *Ba 349*. These rocket boosters were heavy. Each 533 weighed 187 pounds [85 kilograms]. Attachment gear/explosive bolts weighed another 22 pounds [10 kilograms] each. The diglycol-dinitrate solid propellant weighed 88 pounds [40 kilograms] for a total of 297 pounds each.

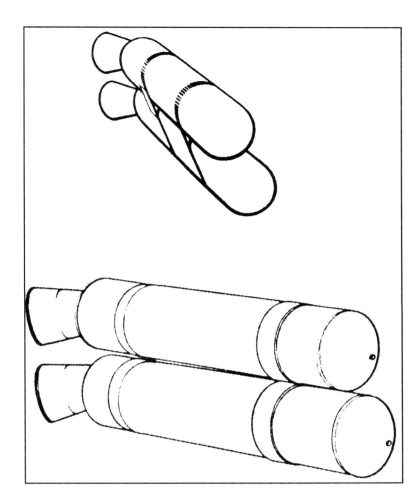

A close up view of a *Ba 349A's* port side *Schmidding 533* booster rocket engines.

The Bachem-Werke Ba 349 "Natter"

Another view of a *Ba 349A*'s port side *Schmidding 533* booster rocket engines.

The rear end of what appears to be a *Ba 349A* "*Natter*" under construction at *Bachem Werke*, Waldsee/Württemberg. Notice that this machine has 4x*Schmidding 533* booster rockets installed and see how their exhaust cones angle outward.

A rare photo featuring a *Ba 349V* with a single *Schmidding 533* booster rocket of 1,102 pounds thrust on its port side...with another one on its starboard side. These *533's* did not have their thrust cones angled outward but parallel to the fuselage's length axis. Notice that the thrust cone has been lined with a semi-circle metal protective plate on the fuselage to keep it from catching on fire. *Bachem* initially thought that unmanned flight testing could be achieved will only two *533* booster rockets. It wasn't enough power to propel the "*Natter*" out and clear of its metal frame launch tower. Several early versions of the 2x*Schmidding 533* rocket booster "*Natters*" got stuck half way up the launch tower and were destroyed by fire.

A close-up view of a set of *Schmidding 533* booster rockets attached to the port side of a *Ba 349A*. This "*Natter*" is mounted on a 70 foot high fresh cut pine tree pole launcher.

A static testing of the *Ba 349V* as shown above with a single *Schmidding 533* booster rocket on each side of its fuselage. Notice that wooden chocks wrapping around the wing at its wing root holds this early test model from propelling itself out the opposite end of the test shed.

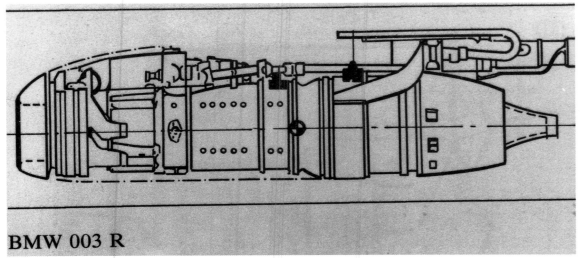

BMW 003 R

A pen and ink drawing of the *BMW 003R* combined turbine jet and rocket-drive engine. During the post war interrogation of *Willy Fiedler* he stated that *Erich Bachem* was considering mounting this combined engine in the fuselage of a *Ba 349A or 349B* (this was confirmed by *Hans Zübert* to this author). *Willy Fielder* did not elaborate on this proposed design modification to the American interrogators so it is not exactly known today how well a "*Natter*" would lift itself off the ground in pursuit of Allied bomber formations. The thrust provided by the *BMW 003R* itself would be no where sufficient to lift it off vertically. Thus a *349A or B* so equipped with the *003R* would have to have wheels and a runway. It has been shown that *Me 262's* were pretty much shot up and destroyed by *P-47s* and *P-51s* they attempted takeoffs and landings. We can assume the same would have occurred to any *003R* equipped *Ba 349As*. But then a "*Natter*" so equipped might still be shot off vertically were it to also have 4x*Schmidding 533* booster rocket engines.

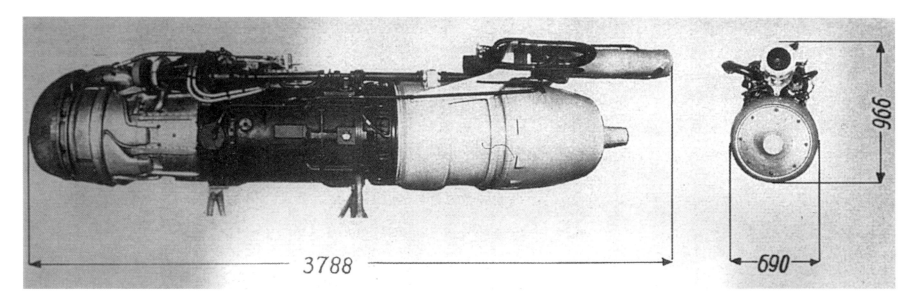

Two views of the BMW 003R: port side and rear view looking forward. The 003R was basically a standard 003A of 1,700 pounds thrust with the addition of a BMW 718 bi-fuel liquid rocket producing a thrust of 2,700 pounds for a period of three minutes. BMW designed the combination engine for aircraft interceptors requiring immense power for climbing in pursuit of high-flying Allied bombers. The HWK 509 produced 3,750 pounds[1,700 kilograms] thrust plus the 4xSchmidding booster rockets produced another 4,408 pounds (combined) [2,000 kilograms combined] thrust for twelve seconds. Thus for twelve seconds the 4,920 lift-off weight "Natter" was propelled sky-ward with 8,158 pounds of thrust. Now a "Natter" could well lift off the ground with a single BMW 003R with its combined thrust of 4,400 pounds plus for three minutes with 4xSchmidding 533 booster rockets putting out their combined total 4,408 pounds of thrust for twelve seconds.

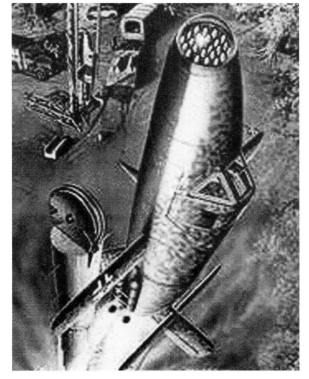

A pen and ink illustration of a Ba 349A "Natter" being launched from its 70 foot high pine tree pole. The pine tree launching device was how Erich Bachem envisioned his wooden interceptor was to be lifted skyward a intruding B-17 bomber formations from mobile sites all over Germany.

This is pretty much how a *Ba 349A* was going to bring down one or more Allied bombers. The "*Natter*" would unlash all of its 24 *R4M 73* mm rocket shells with their proximity fuses right into the bomber formation.

A poor quality pen and ink illustration of a *73* mm *R4M* rocket attack by *Me 110's* on a formation of *B-17* heavy bombers over Germany. *Erich Bachem* believed that his *349A* would bring about the same results and this is how he sold the idea to *Reichsführer-SS Heinrich Himmler.*

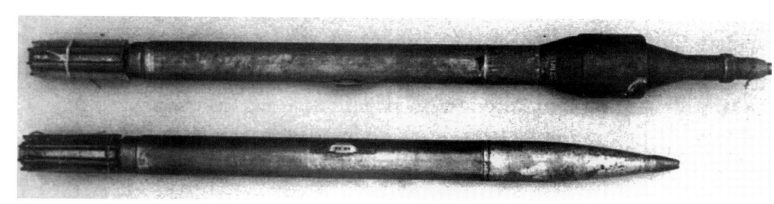

The individual weight of a *73 mm R4M* rocket shell was 8 3/4 pounds [4 kilograms].

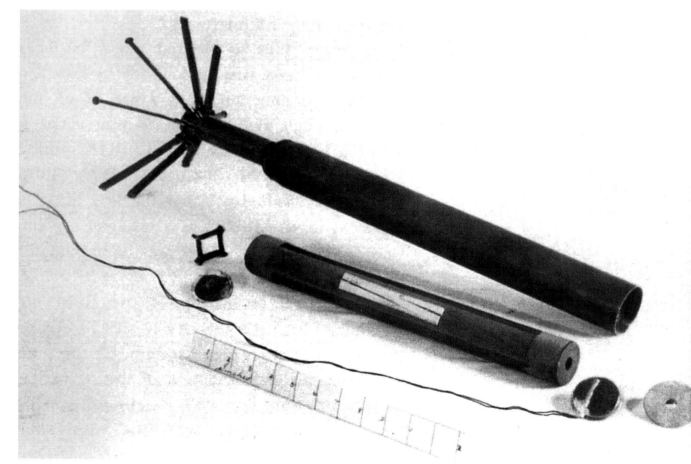

After a *73 mm R4M's* release its eight stabilizing fins sprung open. Each *R4M* carried 500 grams of Hexogen explosive (tetra methylene-trinitramine). Approximately 20,000 *R4Ms* were manufactured by the *Kratzau Werke* in the Sudetenland of Czechoslovakia before the factory and its store of new *R4Ms* were overtaken by the Red Army. It is estimated that about only 1,000 *R4Ms* were delivered to the *Luftwaffe*...the remaining 19,000 lost to the Soviets.

An under wing rack of *73* mm *R4M* rocket shells installed on the starboard wing of a *Me 262*.

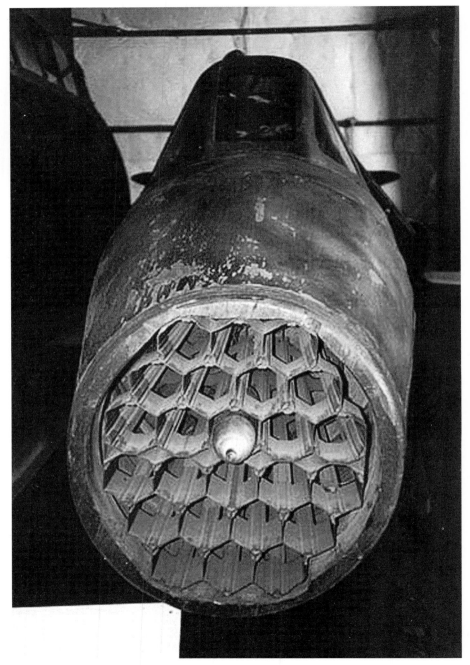

The National Space and Aeronautics Museum's *T2-1011*(*Ba 349A*) in storage at Silver Hill, Maryland. Only one solitary *R4M* rocket shell remains in its rusting *73* mm hexagon rocket rack.

The nose of the *BA349* was pretty much filled up with its 24 *R4M* rocket shells. This "*Natter*" has been photographed at Freeman Field post war. The *R4M* was initially meant for the *Me 110* only, but later, after its unexpected successes, it was released for general use on *Bf 109s*, the *Fw 190*, and then for use by all *Luftwaffe* aircraft including the *Me 262*. The success of the *R4M* exceeded all expectations. During a first trial attack in Spring 1945, the whole attack lasting only a few minutes, six *R4M-carrying* Me *262s* claimed to have shot down fifteen *B-17E* bombers out of an attacking formation at a range of about one mile [1,700 meters] without loss to themselves. Later, in the closing days of the war, it was reported that in April 1945, twenty-four *Fw 190s*, again without loss to themselves, using *R4M* rocket shells destroyed 40 Allied bombers.

A *Ba 349* with its 24 nose-mounted *R4M* rocket shells. This "*Natter*" was on display at Freeman Field post war. Behind the *349A* is a *Junkers Ju 290* "*Alles Kaputt*."

A *Ba 349A* minus is *R4M* rocket tube cover. The individual sheet metal rocket-holding tubes are clearly visible in this nearly completed "*Natter*" at the *Bachem Werke*, Waldsee/Württemberg.

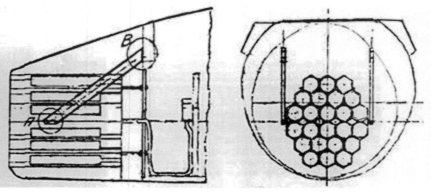

Below: A pen and ink drawing of a *R4M 73* mm rocket shell hexagon rack and its position within the "*Natter's*" nose cone: from the port side (left) and a front on view (right).

Below left: A welded up *53* mm *R4M* rocket shell canister for holding 32 rockets and installed in the nose of a "*Natter*" to check for fit and alignment. It is not known to this author if any "*Natter*" was equipped with the 32 *55* mm *R4M* rocket canister such as the one shown.

Right: A pen and ink 2-view drawing of how the proposed 32 *55* mm *R4M* rocket tubes would appear as installed in a *Ba 349A or B*. Port side (left) and nose (right).

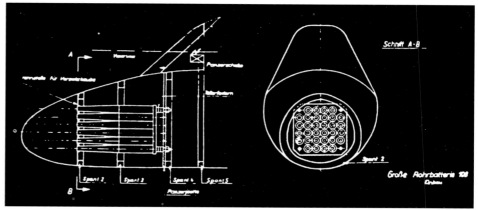

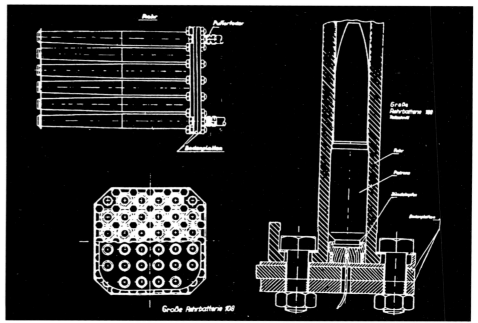

A pen and ink close-up drawing of a 32 *55* mm *R4M* rocket canister plus one of its *55* mm rocket shells.

Field assembly for a highly modified Ba 349. First its cockpit open area is smaller than ones found on the Ba 349A. The bulkhead behind the pilot's head is different, too, being more vertical. Finally, field technician shown appears have removed or is the process of installing the armament pack containing as many as a 46 R4M rocket unit electric firing assembly. Numbers stenciled on the nose section/armament pack are appear to start with 65 32?, however, the last numerals are obstructed by the technician's right hand. It is not known to this author what these numbers mean.

A nice view of the "Natter's" typical 24 73 mm R4M rocket rack mounted in what appears to be a fully assembled but unpainted Ba 349A. This machine has a ring sight near the cockpit windscreen plus a single aiming rod for more accurate aiming by the pilot.

Left: Oberleutnant Lothar Sieber. Many stories circulate about this test pilot's background, such as the allegation that he had been sentenced to a military prison. When approached to do the first manned test flight he accepted...provided he could spend a two-week vacation with his girl friend or wife before the manned test flight. *Willy Fiedler* in a telephone conversation with this author said that what has been written about *Lothar Sieber* is myth. First, *Lothar Sieber* had not been in prison and released to pilot the first manned flight of the "*Natter.*" *Sieber*, said *Fiedler* had lost his officer's rank in the *Luftwaffe* as a result of a court martial proceeding. He was told that he could get his rank back by becoming a test pilot for the *Bachem Werke*. *Fiedler* said that *Sieber* never told anyone why he was court martialed in the first place and no one asked.

Right: Oberleutnant Lothar Sieber. Sieber was an accomplished pilot with a great many flying hours. *Fiedler* told this author that *Sieber* was also a person of considerable daring and this is why *Erich Bachem* wanted him. Prior to being assigned to *Bachem Werke*, *Sieber* had volunteered to fly a *Junkers Ju 52* transport all alone into the Soviet occupied Ukraine to rescue some partisans. *Sieber* took the *Ju 52* in before dawn. The signal was to be a fire burning next to a meadow where it was safe to land. This he did. But as the *Ju 52* rolled to a stop, he was surprised by what appeared to be men from the Red Army. *Sieber* felt at that moment that it was all over. However, the partisans were dressed in the Red Army clothes. *Sieber* felt that he was a very lucky man to have gotten out of the Ukraine alive.

In the lower right corner *Oberleutnant Sieber* in the flight suit is standing next to *Erich Bachem*. It is 28 February 1945 and snow covers the ground. In the background is *Sieber's Ba 349V M-23* which is being made flight ready by *Bachem Werke* workers.

Lothar Sieber is climbing into the cockpit of the *Ba 349V M-23* minutes before his history- making lift-off. Not very many people were present to witness this first manned flight of a "*Natter*" which had been camouflaged in a mottle pattern. Numerals *2+3* had been painted on the upper surface of the wing while numerals *1+4* were painted on the underside.

Left: *Erich Bachem* has climbed up a ladder to see for himself that *Sieber* is seated securely and that the cockpit canopy of the *M-23* is closed and locked in preparation for lift-off.

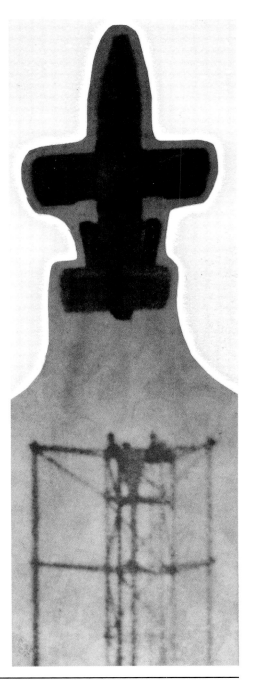

The *M-23* has cleared its launch ramp and *Sieber* was about to experience up to 2.2 times the force of gravity. There is no radio in the *M-23*.

The *M-23* well beyond its metal frame launch tower. On the ground is a huge bank of smoke...products of the *HWK 509* and the 4x*Schmidding 533* booster rockets. *Willy Fielder* told this author that the "*Natter*" rose slowly, without incident, from its tower then did a half roll and continued on its steep climb up.

Right: Pilot *Sieber* and the *M-23* are in trouble. At about 3,000 feet altitude the *M-23* appears to people on the ground that it has turned over on its back and it continues to accelerate...but not at the vertical angle it was suppose to at this height. The small black object below the *M-23* and falling to the ground is its cockpit canopy. *Willy Fiedler* told this author that the "*Natter*" at 3,000 feet altitude was like a black spot in the sky and he and others were hoping that what they were seeing coming down was not the "*Natter*" but a bird instead. We all rushed over a couple small hills to the crash site. When we were on top of the last hill we saw a big hole in the meadow where the "*Natter*" had hit the ground. This was a difficult time for us all, *Fiedler* recalled, and what helped him get through it was the fact that *Erich Bachem* was his friend.

A honor guard from the West German *Luftwaffe* paying tribute to *Lothar Sieber* at his grave site in the village of Heuberg near Stetten. Heuberg honors *Sieber* because he was the first person to make a vertical start in a bi-fuel liquid rocket...the first anywhere in the world, and it happened at Heuberg in early 1945.

Left: This is a reverse angle photo of the *Sieber* and the *M-23* showing that it is out of control. It will crash down while still under power moments later. *Willy Fiedler* was a witness to the loss of *Sieber* and the *M-23*. He told this author that upon *Sieber's* lift-off he pressed his stop-watch. It continue to tick until he saw the "*Natter*" arch over and begin its dive back to earth. At that point he stopped the watch. The time from lift-off to the "*Natter's*" arch over and dive to earth lasted 55 seconds. *Willy Fiedler* told this author that although the cockpit canopy flew open and off he did not believe that the opening of the hinged canopy could have struck *Sieber* either in the back of the neck or head. He believed that when the canopy flew off the open cockpit disturbed the air flow over the machine and this created adverse moments on the "*Natter's*" rudder. As a result, says *Fiedler*, the "*Natter*" may have gone out of control due to this unbalanced air flow and maybe then *Sieber* could do nothing to bring the "*Natter*" back under control, save it, or himself.

Up to ten *Ba 349As* were found at war's end near, including this *Mark II* version, near St. Leonard, Austria. When found it was unpainted, without wood joint sealant, and without camouflage. Notice that no national insignia appears on this machine...this is because *RLM* policy did not allow insignia on any disposable aircraft. The *349A* in this photo is on a simple wagon normally used to move the "*Natter*" around the launch site. It is secured by two web straps as shown in this pen and ink drawing. One strap was placed just aft of the cockpit canopy while the second was just forward of the center fuselage/tail assembly breakpoint.

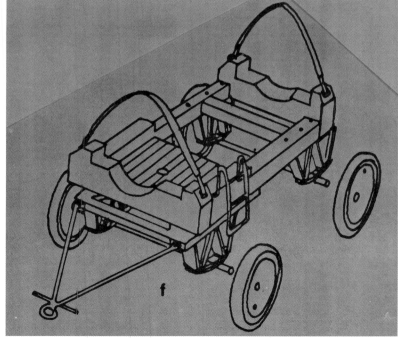

The *Ba 349As* found at St. Leonard, Austria were being transported on simple four-wheeled wooden wagons as well as more heavy duty wagons. *Waffen-SS Oberleutnant Flessner* told this author that near the end of the war he received the order to release his 200 men to the reserve units. Some of his best he kept to help him take three "*Natters*" to the Alps. *Flessner* wasn't given any reason; after all, Germany's unconditional surrender was at hand. Nevertheless, he, several officials from the *Bachem* Werke, and 13 of his hand-picked men transported three "*Natters*" to the top of a valley in Austria and were told to wait for the Americans to arrive. He believes that *Erich Bachem* and others felt that the Americans might want to perfect the "*Natter*" for their upcoming fight with the Russians.

An American soldier (left) with the 44th Division and *Dr.-Philosophy. Heinrich Rieck* (right) 11 May 1945 and standing by the port side of a *Ba 349A* on a mountain top near St. Leonard, Austria behind which, is a stunning view of an Austrian valley. This "*Natter*" has been placed on a long-range heavy duty transporter. *Dr. Rieck* had been a trained political economist who had turned to engineering. He was working on a replacement bi-fuel rocket engine for the *HWK 509* and had been assigned to the *Ba 349A* program by *Waffen-SS General Wolff*. How, where, and when *Dr. Rieck* would have come up with a better bi-fuel rocket engine is difficult to imagine since the *109-509* was a matrue rocket engine in the *Luftwaffe* and being produced by *HWK* cheaply, quickly, and in large numbers.

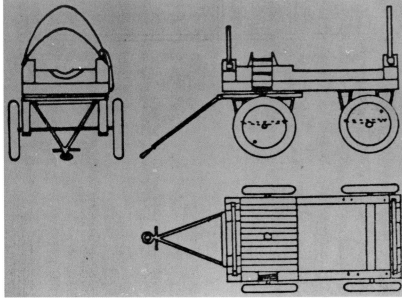

A pen and ink 3-view drawing of the typical launch area transport cart/wagon for an operational *Ba 349A*.

A poor quality photo of *Dr. Rieck* appearing to be describing some feature of a *Ba 349A's* aft fuselage/tail assembly to an American soldier at St. Leonard.

In this poor quality photo *Dr. Rieck* appears to be getting ready to climb into the cockpit of a *Ba 349A* to show his American captors at St. Leonard how it's done.

A poor quality photo of the nose-on view of the "*Natter*" which *Dr. Rieck* has disappeared into as a demonstration for the Americans.

A poor quality photo showing *Dr. Rieck* climbing into the same *Ba 349A* to give the Americans a view as to how it was done by a *Bachem Werke* test pilot.

The object in this photo appears to be the entire upside down nose section from a *Ba 349A*. Rudder foot pedals can be seen at the top of the object along with other miscellaneous cables. This nose section was also found at St. Leonard, Austria.

Another view of one of the ten *Ba 349As* found at a mountain top site near St. Leonard, Austria. The American *GI* is pointing to the exhaust orifice of the *HWK 509's* bi-fuel rocket engine. The nose of another *Ba 349A* is to the immediate right and shows empty *R4M 73* mm rocket rack.

The only known *Ba 349* of the two brought to the United States post war to have survived. This machine has been placed next to a *Junkers Ju 290* and both are on display at Freeman Field, Seymour, Indiana. The *Ba 349A's* dark gray camouflage and *Halkenkreuz* [swastika] are both incorrect and having been applied after its capture. Disposable aircraft such as this *Ba 349A* were not allowed by the *RLM* to carry *Halkenkreuz*. This machine is *T2-1011* and is presently owned by the National Air and Space Museum.

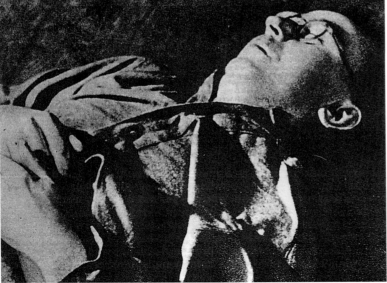

Like the several abandoned and dead *Ba 349A's* found postwar, the body of the "*Natter's*" powerful supporter *Reichsführer-SS Heinrich Himmler* lies dead, too, after swallowing a capsule of cyanide several hours after being arrested by the American military as he and several of his *SS* colleagues attempted to pass through a check point and were detained for questioning. *Himmler* had shaved off his mustache and was wearing a black eye patch to help him look like someone other than the feared and powerful figure he had become. *Himmler* was born 1900 and died by his own hand 23 May 1945 at #031 Civilian Interrogation Camp, Luneburg, Germany.

A starboard side/nose view of *Ba 349A T2-1011* at Freeman Field and featuring its 24 *R4M* rocket tubes. During lift-off these rocket shells would be protected by a jettisonable plexiglass nose-cone.

A *Ba 349A* "*Natter*" on a wood stand beneath the trees. The camouflage on this machine is much different from that found on *Ba 349A T2-1011*. Is this the second "*Natter*" reported by General McDonald to have been delivered to the United States post-war? It probably is.

The USAAF's *Ba 349A* coded *T2-1011* indoors at Freeman Field as seen from its port rear side. Several interesting items are seen: the dark round items on the fuselage directly beneath the tail assembly are the rear attaching points for the *Schmidding 533* booster rockets. The forward attaching points are those two rectangular strips seen aft the wing's trailing edge. The ventral fin is seen with its ventral rudder, at the bottom its launch rail guide, and forward the ventral fin is where the parachute exit hatch is located. The parachute was located on the *349A's* starboard side so it is not visible in this view.

Another photo of what appears to be the second captured *Ba 349A* on semi-trailer. This "*Natter*" was widely displayed throughout the United States post war...unlike *T2-1011*. The sign on the flat-bed trailer reads that it will be on display 3 August (1945) at Douglas Airfield, Santa Monica, California. This was Douglas Aircraft's airfield and now the Museum of Flying occupies part of the old airfield. Although it is hard to tell, the *Ba 349A* in this photo appears to have a different camouflage than the one placed next to the *Ju 290* or *T2-1011* at Freeman Field. The *Ba 349A* in this photo appears to be more of mottle camouflage and, in fact, this paint job appears to be more worn as seen in different areas of the fuselage. Could it be the second *Ba 349A General McDonald* reported he was shipping to the United States and which is presently considered lost?

The Bachem-Werke Ba 349 "Natter" 109

The *T2-1011* and seen from overhead featuring its starboard nose side at Freeman Field. The shiny band painted on the starboard wing was added post war. It was a black band with a red *Halkenkreuz* (*swastika*) appearing inside. Completely non regulation as well as the fuselage's paint job.

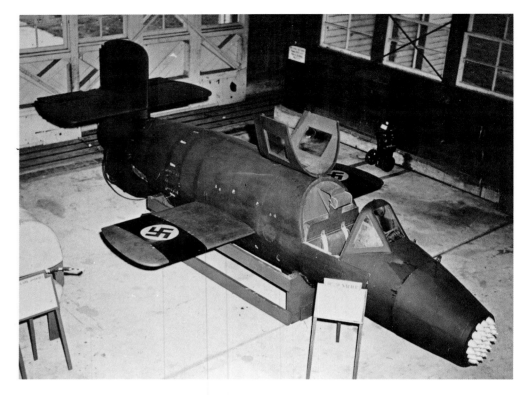

A full port side view of the National Air and Space Museum's *Ba 349A* on display at Freeman Field, Seymour, Indiana post war. It was assigned a Foreign Equipment code number *T2-1011* after its capture.

A close up of Freeman Field's *T2-1011* with its open cockpit canopy. The large round cover appearing just forward of the wing root is the filler cap for the *C-Stoff*. The filler cap for the *T-Stoff* is seen just under the square plexiglass window of the cockpit canopy. When *Oberleutant Lothar Sieber* lost his life in the first manned powered test flight (1 March 1945), the cockpit canopy flew open and off. On this version the head rest was attached to the canopy (not so in this new and improved *T2-1011*) and it is believed that *Sieber's* head was violently pounded against the cockpit's rear bulkhead armor from air rushing in.

The *T2-1011* (*Ba 349A*) as seen from its port side nose. To the upper left corner of the is a white tank with a second tank beneath it. This item is the fuel tank which was built into each *HWK 509* powered "*Natter*." It also was on display for civilians. The larger top tank held *T-Stoff* while the smaller tank held *C-Stoff*.

The Bachem-Werke Ba 349 "Natter"

The *T2-1011* sitting in its wooden cradle, strapped down, and with its cockpit canopy closed. Freeman Field, late 1945. It appears that this machine has been made ready for a move to another location...display or storage.

A close up of the *T2-1011* at Freeman Field showing its round ring sight for aiming its 24 *R4M* rocket shells into an Allied bomber formation.

112 The Bachem-Werke Ba 349 "Natter"

Below: The *T2-1011* inside a hangar at Freeman Field showing its full complement of *R4M* rocket shells. Today at the NASM's restoration/storage facility, Silver Hill, Maryland only one rocket shell remains in its nose rocket rack.

This *Ba 349V* was unmanned test vehicle known as *M-17*. It was painted yellow, with the black strips applied as a tracking aid when it was tested.

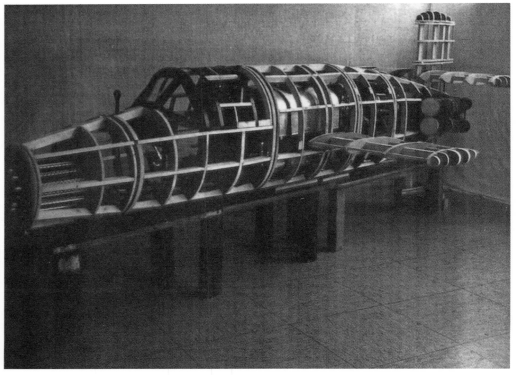

What appears to be a *Ba 349V* without any plywood covering on its fuselage, wings, and tail unit. It appears, too, to have been constructed as a construction aid so that new laborers could gain a visual acquaintance of the machine they were building. Location and date unknown to this author.

Left: This is the second known *Ba 349A* to have survived up to the present time. It is owned by the *Deutsches* Museum, Munich. Although the photo presented is black and white, this *Ba 349A* has been faithfully camouflaged in the colors of the *M-17,* and right down to the stenciled reward information on the tail plane. A very nice restoration. Photographed by Gary Hethcoat.

NASM's *Ba 349A* and known as the *T2-1011* from its post-war evaluation at Freeman Field. Behind the *349A* is the fuselage of a former Japanese *G4M Ishiki Rikukoh* "Betty" bomber. Both machines have been photographed at NASM's Silver Hill Restoration Center, Silver Hill, Maryland.

The Deutsches Museum's M-17 painted Ba 349. The size of the M-17 (small) can be judged by the man standing near its port side Schmidding 533 booster rocket and the port side elevator. Photographed by Gary Hethcoat.

Another view of NASM's T2-1011 in storage at their Silver Hill, Maryland facilities with the Japanese "Betty" in the background. Photographed by Gary Hethcoat.

The *Ba 349A* (*T2-1011*) as it appears today at NASM-Silver Hill. The nose, nose cone and cockpit windscreen have been removed. It comes off in one piece. If you look closely you can see the pilot's control stick just inside the empty cockpit.

Left: A front-on view of a scale model of the metal launch tower built to test the *Ba 349V* early in its development program. Operational "*Natters*" would be launched from sites throughout Germany with the aid of a simple, fresh-cut pine tree pole. Mounted in this metal-frame launcher is *Ba 349V M-13*. Scale model by *Reinhard Roeser*.

Right: The support equipment found around the base of a *Ba 349V's* metal launch tower is shown in this scale model. The two large white box-like items contained wrenching equipment to lift the "*Natter*" up in the tower before it was secured. Scale model by *Reinhard Roeser*.

It appears that a spreader bar with two cables are used to haul the "*Natter*" (a *Ba 349A* shown on the tower) up on the metal frame launching ramp. Scale model by *Reinhard Roeser*.

Left: The "*Natter*" was attached to this near vertical metal frame launching ramp which had three guide rails. According to expert gliding test pilot *Hans Zübert* who was on loan to *Bachem* from *Horten* brothers, the metal launch tower was a relatively complicated piece of equipment about 65.5 feet [20 meters] high. It was rotatable and inclined from 90 degrees to 120 degrees. Scale model by *Reinhard Roeser*.

The *RLM* authorized under-surface color for the "*Natter*" was *White 21*. Scale model by *Reinhard Roeser*.

Left: Uppersurface camouflage of the typical "*Natter*" consisted of *Light Blue 76* with a dense mottle of *Gray-Violet 75*. Scale model by *Reinhard Roeser*.

In this photo a *Ba 349A* is about to be lifted up off its cradle. The tow truck seen in the far right of the photo is being used to pull the cable which wraps around a large diameter pulley at the top of the pine tree and then comes down where it is attached to a spreader bar, each cable attached to the "*Natter*" via removable wooden stocks wrapping around the wing at their wing-root.

Left: **The use of a simple pine tree stripped of all its limbs and bark and secured several yards into the ground, perhaps in concrete, would be the finalized way of launching "*Natters*," up at Allied bomber formations throughout Germany.**

The "*Natter*," as seen from its port side, has been lifted off of its cradle and *Bachem Werke* personnel are man-handling the *349A* to bring it vertical along side the pine pole.

In this view we see the starboard side of the "*Natter*" being lifted up to be secured to the pine tree pole launcher. Underside the starboard wing appears a painted on "dash" to track it visually after lift-off.

A view of the "*Natter*" from its starboard side and behind. For ease of tracking after lift-off, this *349A* has a cross painted on the upper surface of its starboard wing. The port wing has a large, painted on circle.

The "*Natter*" appears to be fully secured to its pine tree pole launching apparatus. On the underside of its port wing *Bachem Werke* personnel have painted what appears to be a triangle...helpful when tracking the machine after lift-off.

The *Ba 349A's* mounted on its cheap and quickly accomplished 70 foot high fresh-cut pine tree pole launching device. All is ready now, and dawn is breaking.

The pine tree pole launching system as seen from several hundred feet away. This author is unaware of any photographs showing a "*Natter*" being flight tested from a pine tree pole launch.

The Bachem-Werke Ba 349 "Natter"

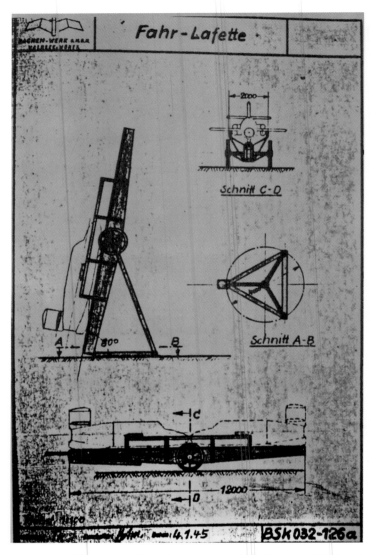

A poor quality pen and ink drawing from the *Bachem Werke* of a proposed highway portable launching device known as the "*Fahr-Lafette*." It appears that two "*Natters*" could be transported on this device. When it came time to fire them one of the "*Natters*" would be removed and the entire device placed on its trailer-towing end as shown in the illustration. This author is not aware if any *349's* were tested fired on the "*Lafette*."

Depicted here is *Oberleutant Lothar Sieber's M-23* on a pine tree pole launcher. *Bachem Werke* personnel have removed the tail assembly in order to service its *HWK 509* bi-fuel liquid rocket engine. Another "*Natter*" is on its wooden cradle and being carried by a 2.5 ton *Henschel* truck. Scale model and photographed by *Jamie Davies*.

The nose and cockpit canopy of *M-23* as seen from the top of its pine tree pole launcher. A second *Ba 349A* and giving a good view of its planform is on the end of a 2 1/2 ton *Henschel* truck. Scale model and photographed by *Jamie Davies*.

The *M-23*, giving a good view of its *HWK 509* thrust tube and combustion chamber, is shown mounted on a pine tree pole launcher. Its tail assembly is on the 2 1/2 ton *Henschel* truck. Notice the cluttered burnt ground around the launch site. Its cluttered with spent *Schmidding 533* solid fuel rocket booster engines. Scale model and photographed by *Jamie Davies*.

The Bachem-Werke Ba 349 "Natter"

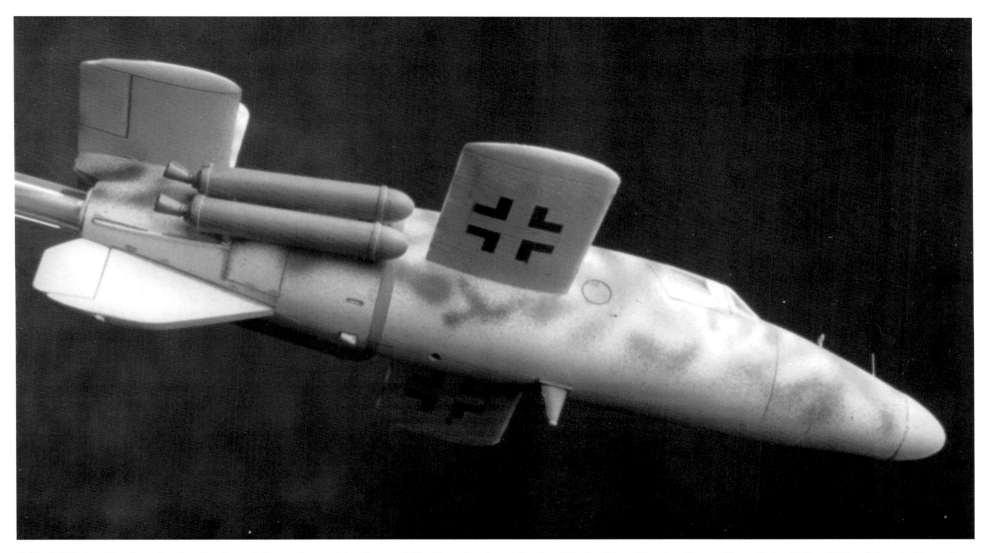

A *Ba 349A* shortly after clearing the launch tower is reaching for an Allied bomber formation's attitude. It is still using its auxiliary *Schmidding 533* solid rocket boosters. About twelve seconds after lift-off these spent boosters were released and fell away. The simplified *B6 Balkenkruez* applied to the wing's under surfaces is outlined in black. Scale model and photographed by *Jamie Davies*.

Opposite: A new appearing *Ba 349A* mounted on its pine tree pole launcher. The *Balkenkruez*, the late war simplified *B6* style outlined in white, has been applied to the upper surfaces of both wings. Scale model and photographed by *Jamie Davies*

The port side rear view of *Ba 349A* in flight as seen from above heading into the sun. Scale model and photographed by *Jamie Davies*.

A *Ba 349A* has leveled off at an altitude which will bring it face to face with an Allied bomber formation. Notice that this "*Natter's*" plexiglass nose cone is still in place but it will soon be jettisoned as it positions itself to fire off the 24 *R4M 73* mm rocket shells. Scale model and photographed by *Jamie Davies*.